Lecture Notes
in Business Information Processing

347

Series Editors

Wil van der Aalst
RWTH Aachen University, Aachen, Germany
John Mylopoulos
University of Trento, Trento, Italy
Michael Rosemann
Queensland University of Technology, Brisbane, QLD, Australia
Michael J. Shaw
University of Illinois, Urbana-Champaign, IL, USA
Clemens Szyperski
Microsoft Research, Redmond, WA, USA

More information about this series at http://www.springer.com/series/7911

Tim A. Majchrzak · Cristian Mateos ·
Francesco Poggi · Tor-Morten Grønli (Eds.)

Towards Integrated Web, Mobile, and IoT Technology

Selected and Revised Papers from the
Web Technologies Track at SAC 2017 and SAC 2018
and the Software Development for Mobile Devices,
Wearables, and the IoT Minitrack at HICSS 2018

Springer

Editors
Tim A. Majchrzak ⓘ
Department of Information Systems
University of Agder
Kristiansand, Norway

Cristian Mateos ⓘ
ISISTAN-UNICEN-CONICET
Tandil, Argentina

Francesco Poggi ⓘ
University of Bologna
Bologna, Italy

Tor-Morten Grønli ⓘ
Kristiania University College
Oslo, Norway

ISSN 1865-1348 ISSN 1865-1356 (electronic)
Lecture Notes in Business Information Processing
ISBN 978-3-030-28429-9 ISBN 978-3-030-28430-5 (eBook)
https://doi.org/10.1007/978-3-030-28430-5

This Springer imprint is published by the registered company Springer Nature Switzerland AG
The registered company address is: Gewerbestrasse 11, 6330 Cham, Switzerland

Towards Integrated Web, Mobile, and IoT Technology: Foreword

The World Wide Web is relentlessly evolving. Once it was a single interconnection of static, physically distributed content passively accessed by human users through personal computers. During the explosion of Web-based social networks, the Web evolved into an environment allowing users worldwide to interact and collaborate in the creation of user-generated content within many virtual communities. In this line, Web 2.0 is the umbrella term used to encompass several developments which followed, namely social networking sites and social media sites (e.g. Facebook), blogs, wikis, folksonomies (e.g. Flickr), video sharing sites (e.g. YouTube), collaborative platforms, and mashup applications. Many technologies such as HTML5, CSS3, AJAX, and client-side scripting helped to bring these ideas into practice. Simultaneously, the mobile ecosystem evolved and through the smartphone revolution from 2007 mobile applications ("apps") became the predominant factor on our personal devices.

The current Web is an evolutionary step from the Web 2.0 in that access to content is nowadays ubiquitous, content itself is far more heterogeneous, and "users" come in mixed and different flavors. First, ubiquitous access has been mainly pushed by the inception of mobile computing and mobile devices; in fact, reports show that by 2020 the number of mobile device users will be about 70% of the global population. This has given birth to new areas of research beyond Web that seek to produce tailored computing paradigms – namely Dew, Fog, and Edge Computing – aimed at both addressing mobile devices' inherent limitations and exploiting their capabilities to build novel applications. Secondly, served and published Web content is not only those following traditional interchange formats (text, images, video) but also executable code or Web APIs (e.g. Mashape.com, ProgrammableWeb.com), from which new applications can be built and in turn published back to the Web. The recent notion of "Web of objects," which find its roots in Web-accessible IoT applications, promotes the interconnection of hardware elements capable of producing huge amounts of sensor data. Finally, the role of Web application end users and Web developer and designers is somewhat blurry, due to modern Web technologies that greatly simplify the creation and deployment of rich Web sites that might consume Web-accessible services. In addition, the advent of Semantic Web technologies paves the way for one research stream towards the creation of intelligent applications, and thus the tandem human user-browser is no longer the only way to take advantage of Web content.

In this context, novel approaches and techniques, new tools and frameworks are needed to address the increasing complexity of the distributed computing paradigms that are coming and the applications therein. Motivated by these necessities, we have therefore brooded over the idea of putting selected, extended papers from pertinent highly-reputed conferences into a collection with a fitting theme. This volume contains extended papers from (a) the Web Technologies track at the 33rd ACM/SIGAPP Symposium On Applied Computing (ranked B according to core 2018), (b) the Web

Technologies track at the 32nd ACM/SIGAPP Symposium On Applied Computing (ranked B according to core 2018), and (c) the Software Development for Mobile Devices, Wearables, and the Internet-of-Things Minitrack at the 51st Hawaii International Conference on System Sciences (ranked A according to core 2018). Thus, it contributes with a uniform view of cutting-edge research in Web, Mobile, and IoT technologies, having these venues as a primary source.

The first contribution in this volume is "Metamorphic Testing of Mapping Software," by Joshua Brow, Zhi Quan Zhou, and Yang-Wai Chow. The authors apply the concept of metamorphic testing to Web-based Map services, namely the wide-spread Google Maps and OpenStreetMap. Their work deepens the understanding of this technique in the context of the Web, as well as provides a concrete evaluation.

The second article, "Multi-criteria Recommendations by Using Criteria Preferences as Contexts," by Yong Zheng, addresses recommender systems. This classical topic in the realm of Web technologies is broadened to multi-criteria recommender systems (MCRS). The author presents exhaustive work on the advancement of the state of the art.

In the third article, "Towards Pluri-Platform Development: Evaluating a Graphical Model-Driven Approach to App Development Across Device Classes," by Christoph Rieger and Herbert Kuchen, the authors focus on cross-platform app development. However, due to the complexity of developing apps not for 'merely' incompatible platforms on similar devices but rather among the devices classes, they propose the term 'pluri-platform development'. The authors present work on MAML, a framework they suggest as a step towards pluri-platform development.

The fourth article is "What Matters for Chatbots? Analyzing Quality Measures for Facebook Messenger's 100 Most Popular Chatbots," by Juanan Pereira and Oscar Díaz. The authors chose a very timely topic, which will likely gain even more momentum. Their work indicates that the most popular chatbots that they analyzed were rather simple – an observation that they expect to change within a short time.

In the firth article, "Linguistic Abstractions for Interoperability of IoT Platforms," by Maurizio Gabbrielli, Saverio Giallorenzo, Ivan Lanese, and Stefano Pio Zingaro, we move to the Internet-of-Things. The authors address IoT interoperability – a major challenge of IoT application. Building on Jolie, the authors propose a language-based approach for the integration of disparate IoT platforms.

The sixth and final article is "Energy-Efficient Scheduling of Tasks with Conditional Precedence Constraints on MPSoCs," by Umair Ullah Tariq, Hui Wu, and Suhaimi-Abd-Ishak. As the most formal paper in this volume, it targets energy-efficiency, a topic particularly relevant in mobile computing and embedded systems. The authors present a heuristic-based approach that significantly improves the state of the art.

We sincerely thank the anonymous reviewers that helped us to review the papers included in this volume. It is worth mentioning that the quality of the contributions presented in this volume is also due to the hard work of the members of the Program Committees of the Web Technologies Track within SAC (32nd and 33rd editions) and the Software Development for Mobile Devices, Wearables, and the Internet-of-Things Minitrack within HICSS (51st edition). Finally, we would like to thank Springer for making this volume possible.

We wish you a pleasant and stimulating read.

Tim
Cristian
Francesco
Tor-Morten

Contents

Metamorphic Testing of Mapping Software

Joshua Brown, Zhi Quan Zhou$^{(\boxtimes)}$, and Yang-Wai Chow

Institute of Cybersecurity and Cryptology, School of Computing
and Information Technology, University of Wollongong,
Wollongong, NSW 2522, Australia
jb740@uowmail.edu.au, {zhiquan,caseyc}@uow.edu.au

Abstract. Mapping software is difficult to test because it is very costly to evaluate its output. This difficulty is generally known as the oracle problem, a fundamental challenge in software testing. In this paper, we propose a metamorphic testing strategy to alleviate the oracle problem in testing mapping software. We first conduct a case study to test Google Maps, the most popular web mapping service. The results of the case study show that our testing approach is effective, with the detection of several real-life bugs that can hardly be exposed under conventional testing paradigms. Following this, we conduct an analysis of the system OpenStreetMap, well-known open-source mapping software built and maintained by a community of users. We show the potential of metamorphic testing for such systems. These case studies show that metamorphic testing can be applied to mapping software for both verification and validation purposes.

Keywords: Mapping software · Google Maps · OpenStreetMap · Navigation software · Web service · Graphical User Interface · Software testing · Oracle problem · Metamorphic testing

1 Introduction

A map is a representation of the world. It is a vast and complex system of interconnecting land and sea, physical features, roads, intersections and other features which is further extended through the use of attributes to refine the maps representation. In the world, there are over 64 million kilometres of road [2], which continues to grow and change every day. The mapping software has a routing utility which is designed to plan an optimal route between two points within the constraints it has been passed; the route contains the process that the user should follow to reach the destination.

Mapping systems are one of the most popular applications on the Internet, and in particular on smart phones. They are the most popular application

An initial version of this paper was presented at HICSS-51 [1].

© Springer Nature Switzerland AG 2019
T. A. Majchrzak et al. (Eds.): *Towards Integrated Web, Mobile,
and IoT Technology*, LNBIP 347, pp. 1–20, 2019.
https://doi.org/10.1007/978-3-030-28430-5_1

installed on over 50% of the global smart phone market [3]. Furthermore, these systems are mission critical applications, as their failure could potentially cause traffic accidents, especially when they are used for the navigation of autonomous vehicles such as self-driving cars. Mapping software and its components, therefore, must be thoroughly verified and validated.

In order to verify and validate software systems, testing is essential. It is widely accepted that, in a typical commercial software development project, the cost of testing can easily exceed 50% of the total development budget. Testing involves executing the *software under test* (SUT) with a set of test cases together with a mechanism against which the tester can decide whether the outcomes of test case executions are correct (that is, a *test oracle*). The *oracle problem* refers to the situation where an oracle does not exist or it is theoretically available but practically too expensive to be applied. The oracle problem is a fundamental challenge in software testing practice but this problem is often ignored by the research community—compared with many other aspects of testing such as automated test case generation, the challenge of test oracle automation "has received significantly less attention, and remains comparatively less well-solved" [8].

There is a severe oracle problem when testing some of the critical features of mapping software. For example, the real-world road networks are so complex that in most situations it is infeasible to validate whether or not a route returned by the mapping software is correct and optimal, except trivial cases. This difficulty results in a scenario where developers cannot utilize conventional testing techniques. In fact, literature on automated testing of mapping software is very limited.

A growing body of research has investigated the concept of *metamorphic testing* (MT) [10–19,26,31], and has proven MT to be a highly effective testing paradigm for the detection of real-life software faults in the absence of an ideal test oracle. The idea of MT is simple: instead of focusing on the correctness of each individual output, MT looks at the relations among the inputs and outputs of *multiple* executions of the SUT. Such relations are called *metamorphic relations* (MRs), and are necessary properties of the intended software's functionality. For example, the accuracy and completeness of the search results returned by a search engine is difficult to assess [14,34]. Nevertheless, MT can be applied by identifying the following metamorphic relation: a stable search engine should return similar results for similar queries. For instance, although a search for [today's movies in Honolulu] and a search for [Honolulu movies today] may return different results, the two sets of search results should have a large intersection, if the search engine under test is robust [14]. Because MT looks at the relations among multiple SUT executions instead of focusing on the verification of each individual output, it can be performed in the absence of an ideal oracle, alleviating the oracle problem.

Although the basic concept of MT is simple, it requires specific study when applied to different application domains [15]. This is because different application areas can have different properties of interest to investigate. The present paper proposes applying MT to test mapping software in order to alleviate the oracle problem in testing such systems. We conducted a case study with Google Maps to test its mobile applications, its web service APIs (namely, the Directions API), as well as its Graphical User Interface (GUI) at maps.google.com. Google Maps was selected for the case study because it was the most popular mapping system [3] (except in China, where Google services could not be accessed).

In this study we ask the following research question:

RQ: Can we have a practical and effective method of automatically testing mapping software in the face of the severe oracle problem?

Following this case study we provide a plan to extend the testing methodology to OpenStreetMap. OpenStreetMap was selected because it was the most popular open source mapping software. In addition, it had a large base consisting of over 4.7 million users, 4.6 billion nodes, an average of three million changesets per day and it had over one million contributors [4,5]. OpenStreetMap had some prominent users including Baidu Maps, Uber, SnapChat, the White House, foursquare and Wolfram Alpha [6,7].

The contributions of this paper are summarized as follows:

- This work shows a proposal of the use of metamorphic testing to address the *automated* testing of mapping software.
- This work shows a case study of testing Google Maps from a user's perspective.
- This work shows the detection of real-life bugs in Google Maps, and demonstrates the effectiveness of metamorphic testing.
- This work shows a preliminary analysis of testing OpenStreetMap from a developer's perspective.
- Given the limited literature on automated testing of mapping software, the study of the functional correctness of driving navigation (which is an essential feature of mapping software) presented in this paper, together with the detection of real-life bugs, is a significant piece of pioneering work in engineering mapping software.

The testing methodology presented in this study can also be applied to other forms of mapping systems beyond Google Maps and OpenStreetMap, and can be used as the foundation for future work in the verification and validation of systems such as self-driving cars [33] and delivery robots [32].

The rest of this paper is organized as follows: Sect. 2 further introduces some background knowledge and Sect. 3 explains our testing approach with a focus on the identified MRs for the mapping software under test. Section 4 presents our test results by highlighting the detected defects in Google Maps. Section 5 makes further discussions and concludes the paper.

2 Background

2.1 Difficulties in Testing Mapping Software

As explained in Sect. 1, mapping software is difficult to test owing to the oracle problem. Worse, the lack of system specifications adds further difficulties to user validation: the vast majority of users do not have access to the detailed algorithm designs and system/subsystem specifications of the mapping software they are using. Without access to these specifications, the user manual or online help pages are the only type of information source available to the users. However, as pointed out by Zhou et al. [14], user manuals or online help pages are usually very brief and are not equivalent to the system specification defined as "an adequate guide to building a product that will fulfill its goals" [20]. Consequently, it is basically impossible for users to evaluate the mapping system they are using against its technical specifications or the intended algorithms.

Zhou et al. [14] pointed out that the above phenomenon can be quite common when testing many types of software applications such as web services, poorly evolved software, and open source software, and showed that MT can be an effective approach to addressing these difficulties caused by a lack of detailed knowledge about the system design and specifications coupled with the oracle problem.

For developers of the mapping software, even if they have the complete documentation and specifications, it is still very challenging for them to test such systems because of the complexity of the underlying algorithms and data. As will be shown in this paper, MT can be an effective testing approach for both developers and users, as well as third-party independent testers.

2.2   Metamorphic Testing (MT)

MT [10,26] alleviates the oracle problem by testing the SUT against prescribed MRs, which are necessary properties of the intended program's behavior. (Readers are referred to Chen et al. [31] for a formal definition of MRs.) The difference between MRs and other types of program properties is that an MR involves *multiple* executions of the target program. Even if the correctness of an individual output cannot be verified due to the lack of an oracle, the tester can still check whether the expected relation among multiple executions is satisfied. If the MR is *violated* for any of the test cases, a *fault* is detected. MT has been applied to many different application domains such as web services and applications, computer graphics, simulation and modeling, machine learning, bioinformatics, decision support, components, compilers, numerical programs, and so on [15]. Recently, MT has been used in combination with fuzzing to detect previously unknown fatal defects in real-life self-driving cars [33].

To illustrate the concept of MT, consider the following example: let $p(G, x, y)$ be a program that calculates the shortest path from node x to node y in an undirected graph G. When G is large and complex, it can be difficult to verify the output of p because no oracle can be practically applied. To perform MT in this situation, we can identify many MRs for the shortest path problem, one of which can state that swapping the origin and destination nodes should not affect the length of the calculated shortest path [21]. Using this MR, a metamorphic test will run p twice, namely, a *source execution* on a *source input* (G_1, x_1, y_1) to produce a *source output*, and a *follow-up execution* on a *follow-up input* (G_2, x_2, y_2) to produce a *follow-up output*, where $G_2 = G_1$, $x_2 = y_1$, and

$y_2 = x_1$. Any violation of this MR (that is, if the source output and the follow-up output are found to have different lengths for some test case(s)) will reveal a fault in p. Many other MRs can also be identified and used to test p [21], and different MRs often have different fault-detection capabilities.

When MT was first proposed, it was designed as a *verification* technique, where an MR is a necessary property of the intended algorithm or system specification to be implemented. In this situation, a violation of the MR reveals a fault in the implementation. Later, it was observed that MRs can also be defined based on user expectations "to reflect what they really care about," rather than based on the algorithms or system specifications of the developers—such algorithms and specifications are often unknown to the users anyway [14]. In this way, MT can be used as a user-oriented approach to perform *validation* and other types of *quality assessment* (such as the assessment of *usability* and *functional completeness*), and hence MT has been developed into a unified framework for software verification, validation, and quality assessment [14].

2.3 Related Work

Nisner and Johannessen [23] reported on the failure rates and the failure modes experienced with Standard Positioning Service (SPS) GPS receivers certificated for aviation use. Another related study was conducted by Wright et al. [9], where they manually evaluated the position accuracy and measurement accuracy (driving distance) computed by GPS receivers against a local area road map. Their study faced a limitation of having a tester drive the route out in order to compare with the road map. They did not find unexplained flaws in the system under test.

Elleuch et al. [24] introduced a process of converting raw GPS data to a routable road map, where they presented an architecture that collected GPS data from a large number of vehicles to get the road traces. They generated a Tunisian map network from their database and map-matched it with Google Maps to compare them and generate the missing road data. The same group of authors continued this work [25] and presented a 100 GB database of geographical coordinates collected from GPS receivers in several vehicles. Their database contained many types of information including speed. Based on these data, they generated a road map of Tunisia and then compare it with the roads shown in Google Maps.

While much research has been done with a focus on the accuracy of the map data [30] and implementations of newly proposed methods and algorithms [28,29], Goodman [27] noted that navigation outages due to receiver software issues may pose great threats to the users and that the high cost of meeting strict software quality standards, and the proprietary nature of Global Navigation Satellite System (GNSS) receiver software, "makes it more difficult to ensure quality software for safety-critical applications. Lack of integrator and user insight into GNSS software complicates the integration and test process, leading to cost and schedule issues." Goodman also noted that "a close relationship with the GPS vendor, open communication among team members,

Independent Verification and Validation of source code, and GPS receiver design insight were keys to successful certification of GPS for operational use by the space shuttle" [22].

3   Our Approach for Testing Navigation Software

3.1   The Identified Metamorphic Relations for Navigation Software

In the first part of our study, we investigate the testing tasks of Google Maps using a user-orientated approach by utilizing the concept of MT, as we identify some MRs (MRs 1–5) for mapping software from the users' perspective. This is because the algorithms and detailed system specifications of the SUT are unavailable for reference. In the next part of this study, we consider the open source software OpenStreetMap, where we identify two additional MRs (MRs 6–7) from the developers' perspective because we can access the internal code and data structures of the system, and hence make use of them. However, we did not implement the test driver for OpenSteetMap due to a limited budget of this project (this is because the coding task for the implementation of MRs 6–7 is non-trivial). Nevertheless, it should be noted that the proposal of MRs 6–7 is generally applicable to navigation systems where the data structures of the underlying maps can be edited, and that MRs 1–5 are generally applicable to most navigation systems. These seven MRs are described as follows.

3.1.1   MRSimilar

The first MR is named MRSimilar. Its design is based on the premise that a mapping system should return similar results for similar queries, in a way similar to a search engine [14].

For instance, after a source output (a route) is generated for a source input (an origin and a destination point), we can produce a follow-up input by very slightly changing the origin and/or the destination. Then, in most situations, the follow-up output should be a route having a cost similar to that of the source output. In this paper, the *cost* of a route is in terms of distance or time depending on the user's preference; monetary costs are not considered.

More specifically, let $d(a, b)$ be a function that gives the cost of an optimal route for travel from point a to point b. MRSimilar states that $d(a, b)$ and $d(a', b')$ should be similar if $a \approx a'$ and $b \approx b'$. Here, $x \approx y$ means that x and y are approximately at the same location.

In our experiments with MRSimilar, each source test case (namely, an origin and a destination point) was formed by means of random selection from a set of addresses (to be explained in Sect. 3.2), and the corresponding follow-up test case was produced by adding a tiny amount of distance (e.g., a few millimeters or centimeters on the same road) to the origin and/or destination point. A comparison was then made assessing the difference between the source and the follow-up outputs. An anomaly would be reported if a large amount of difference (e.g., more than ten meters) was detected.

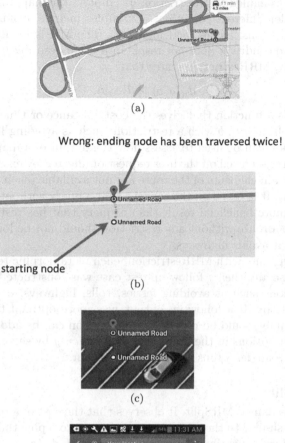

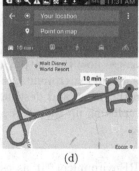

Fig. 1. Google Maps failure detected using MRSimilar with American addresses. (a) The entire driving direction generated by Google Maps (screenshot taken at maps.google.com). (b) Zoom in to show that the destination point has been traversed twice in the route. (c) Further zoom in to show the origin and destination points (d) A similar failure produced by the Google Maps app for Android on a Samsung Galaxy S4 mobile phone.

3.1.2 MRRestriction

The second MR is named MRRestriction. It employs the mapping system's ability to work under different conditions. Examples include avoiding elements of the route such as tolls, ferries, and highways. The MR is focused on ensuring that a restrictive condition does not result in a more desirable/optimal output. More specifically, MRRestriction states that

$$d_R(a,b) \geq d(a,b),$$

where $d_R(a,b)$ is a function that gives the cost (distance or time) of an optimal route for travel from a to b with a restriction, such as avoiding highways.

MRRestriction can be used to assess how the output of a mapping system is affected by the rules placed on the user request or affected by external conditions under which certain elements of the route are not available (such as outside ferry operating hours). It is based on the concept that a query without any restriction should yield a more beneficial result than a query that has restrictive rules on it. For example, a route without any restriction should not be longer and slower than a route that avoids highways.

In the experiments with MRRestriction, each metamorphic test started with a source test case and then a follow-up test case was constructed with the addition of restrictions such as avoiding ferries, tolls, highways, or a combination of these restrictions. If a follow-up output was more optimal than the source output, an anomaly would be reported. A restriction can be added *explicitly* by selecting certain options in the user query or *implicitly* by setting a travel time outside certain road/ferry/bus/train operating hours.

3.1.3 MRSplit

The third MR is named MRSplit. It observes that the cost of a route from a to c via b should be similar to the cost of a route from a to b plus the cost of a route from b to c. More generally, MRSplit requires that $d_m(a_1, a_2, \ldots, a_n)$ should be similar to $d(a_1, a_2) + d(a_2, a_3) + \ldots + d(a_{n-1}, a_n)$, where $d_m(a_1, a_2, \ldots, a_n)$ denotes the cost of an optimal route for travel from a_1 to a_n via $a_2, a_3, \ldots, a_{n-1}$.

MRSplit can be used to assess how the output of the mapping software is affected by intermediate nodes. In our experiments, each source test case included some waypoint nodes between the origin and destination, and a series of follow-up test cases were formed by splitting up the source test case.

3.1.4 MREnvironment

Our fourth MR is named MREnvironment. It assesses how the mapping software's behavior is affected by different user environments. An example of this MR is the same request issued using the API (source input) and the mobile application (follow-up input)—an ideal mapping system should return similar results across these different user environments.

More specifically, let P and Q be two different environments or platforms. Let d_P and d_Q be the cost functions for environments P and Q, respectively. MREnvironment states that $d_P(a,b)$ and $d_Q(a,b)$ should be similar.

In our experiments with MREnvironment, each metamorphic test consisted of source and a follow-up test cases involving exactly the same query but different environments. An anomaly would be reported if the outputs were significantly different (e.g., having more than ten meters difference).

3.1.5   MRFlip

The fifth MR is named MRFlip. It observes that the cost of a route from a to b should be similar to the cost of a route from b to a. More specifically, let $d(a, b)$ be a function that gives the cost of an optimal route for travel from point a to point b. MRFlip states that $d(a, b)$ and $d(b', a')$ should be similar if $a = a'$ and $b = b'$.

For instance, after a source output (a route) is generated for a source input (an origin and a destination point), we can produce a follow-up input by swapping the origin with the destination. Then, in most situations, the follow-up output should be a route having a cost similar to that of the source output.

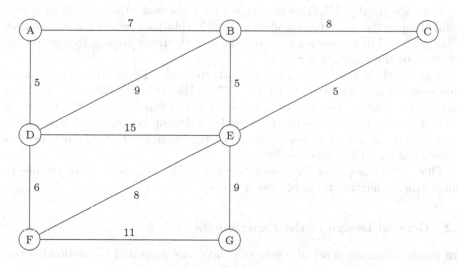

Fig. 2. Sample graph.

It should be noted that this MR may not necessarily be valid if one-way traffic is involved in a route. In our experiments with MRFlip, therefore, violations involving one-way traffic were treated through a separate procedure. For all valid tests, a violation would be reported if a large amount of difference between the source and follow-up outputs was detected.

3.1.6   MRTrimOn

Our sixth MR is named MRTrimOn. It assesses how the mapping software's behavior is affected by changes in the map environments on the route.

It observes that the cost of a source output from a to b which goes through nodes c should be equivalent to or better than the cost of a follow-up route of a to b where c is inaccessible.

An example of this MR can be demonstrated using Fig. 2. Let $d(A, G)$ be a function that gives the cost of an optimal route for travel from point A to point G. Suppose the source output is (A, B, E, G). For a follow-up input where point E in the graph becomes unusable (blocked), the follow-up output should not have a cost that is better (smaller) than the previous cost; otherwise a violation of the MR is reported.

This MR is designed for OpenStreetMap only because we are unable to change the underlying maps of Google Maps.

3.1.7   MRTrimOff

Our seventh MR is named MRTrimOff. It states that the cost of an optimal route returned by a mapping software should remain the same if there is an irrelevant change in the map environment.

More specifically, MRTrimOff requires that the cost of a route from a to b should remain the same if e is made inaccessible (blocked), where e is a node that is not included in the source output (which is the route from a to b originally returned by the mapping software).

An example of this MR can be demonstrated using Fig. 2. Suppose the mapping software returns the route (A, B, E, G) as the optimal route for travel from point A to point G. Because C is not included in this route, in the follow-up input if we block C in the graph then the follow-up output must not return a route that has a better (lower) cost than the route (A, B, E, G); otherwise a violation of the MR is reported.

This MR is designed for OpenStreetMap only because we are unable to change the underlying maps of Google Maps.

3.2   General Design of the Experiments

For the experiments, a set of source test data was generated by randomly sampling 10,000 addresses in Australia and 500 addresses in America, as it was more convenient for the authors to validate Australian addresses.

An individual address formed the basis of a node. Each request to the mapping software can be broken down into five elements: a starting node, an ending node, waypoint node(s) (that is, the intermediate stop(s)), time (departure or arrival time), and restrictions.

3.3   General Design of the Google Maps Experiments

To ensure a stable and consistent testing environment, the real-time traffic feature of Google Maps was turned off during the experiments, as otherwise routing might be affected by live traffic and the test results might not be repeatable. To avoid personalized results, the tester did not log into any online accounts including Google accounts. Furthermore, when an anomaly was observed, the test was immediately repeated. An anomaly would be reported only if it could be reproduced. This treatment was to ensure that the reported anomalies were not caused by the update or dynamics of the maps or algorithms.

The testing process was largely automated by means of test scripts and test drivers. In this research, a total of 1,000 h of test executions were completed across the different environments of Google Maps. As all of the tests were web-based, the impact of hardware selection was very small.

3.4 General Design of the OpenStreetMap Testing Environment

To ensure a stable and consistent testing environment, a clone of OpenStreetMap database and associated systems was compiled and this was to be used in a local system environment to ensure that the results or the system will not be affected by any online changes and allowing the test result set to be repeatable.

Any database changes were made in their own unique separate container to ensure that it would not affect any other test executions. To ensure that hardware or system changes do not affect the experiment, Docker containers will be used to ensure a stable environment across all test executions. This testing process will be largely automated by means of test scripts and test drivers.

We did not continue to implement the test driver for OpenStreetMap because, as explained earlier, the programming tasks for manipulating the underlying maps for MRs 6–7 is non-trivial. Nevertheless, our feasibility study (a "dirty implementation") has proven that MRs 6–7 can indeed be implemented for OpenStreetMap and that it is only a matter of time to turn the "dirty implementation" (proof of concept) into a decent testing tool.

4 Issues Detected in Google Maps

Overall, Google Maps passed the majority of the executed tests; however, there were cases where the system resulted in an unexpected output which was detected by MR violations. A manual inspection of the MR violations revealed several defects in Google Maps. This section will highlight some notable examples of these defects.

4.1 Defects Detected by MRSimilar

In this subsection, we report two failures detected by MRSimilar, where the first failure was detected when searching for a route in America and the second failure was detected when searching for a route in Australia. In software testing, a *failure* refers to erroneous behavior of the SUT.

4.1.1   Searching for a Route in America

When Google Maps was tested against MRSimilar, a source test case yielded an output that had an expected time duration of 1 min and a distance of 0.0 miles. The follow-up test case featured the same starting node and the same conditions and only modified the ending node by a distance of 0.98 cm; the follow-up output, however, suddenly changed dramatically with an expected time duration of 11 min and a distance of 4.3 miles, as shown in Fig. 1.

Figure 1(a) shows the entire driving direction generated by Google Maps. The screenshot was taken by using Google Chrome to access the website maps.google.com. Figure 1(b) zooms in to show a portion of the route surrounding the origin and destination points. It is surprising to see that the destination point has been traversed twice in the driving direction, which is an obvious error. Figure 1(c) zooms in further to show a Satellite View of the exact origin and destination points. Figure 1(d) is an excerpt of a screenshot taken from a Samsung Galaxy S4 mobile phone running the Google Maps app for Android. This screenshot shows that a similar failure was produced using the mobile device.

Given the extremely small distance between the origin and destination points shown in Fig. 1, a normal driver would probably not require any mapping software to guide him or her. It is, however, not the case for self-driving vehicles or robots because such autonomous machinery will always rely on software systems to navigate. It is not acceptable for a driverless car to travel 4.3 miles to reach a destination that is actually only two meters ahead.

4.1.2   Searching for a Route in Australia

Figure 3 shows another failure detected by MRSimilar, using Australian addresses. Figure 3(a) shows a screenshot taken from a Huawei mobile phone running the Google Maps app for Android. It shows that Google Maps returned 1 min walking distance from the current location to "Unit 9/890 Bourke Street, Zetland NSW."

To conduct MT, we also queried the Google Maps mobile app using a slightly modified destination address, which differed from the previous address only in the unit number (namely, using "Unit 8" instead of "Unit 9," as we verified that "Unit 8" was a valid address), but Google Maps returned a very different location that was 2 km away, which required 11 min driving (Fig. 3(b)) or 26 min walking (Fig. 3(c)). A violation of MRSimilar was reported because the difference between the "Unit 8" route and the "Unit 9" route was too large given that they were at the same street address.

After the MR violation was reported, we manually analyzed the test results to investigate the root cause of the failure. We first validated that "Unit 8/890 Bourke Street, Zetland NSW" was indeed a correct address and that Unit 8 and the other units were physically located near each other at 890 Bourke Street. Figure 4 shows a picture taken at the entrance to Unit 8 during the site visit. We further found that Google could not actually locate this unit (although it had been a valid address for a long time) and therefore *automatically* changed the user query from "Unit 8/890 Bourke Street, Zetland NSW" to "Bourke Street,

Redfern, NSW" without explicitly requesting the user to confirm the change. In this modified address, the unit number "8" and the street number "890" were removed, and the suburb "Zetland" was changed to "Redfern." This explains why the location returned by the Google Maps app was 2 km away.

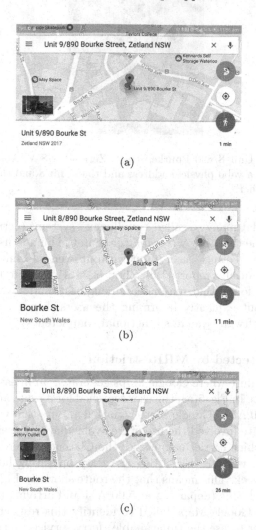

Fig. 3. A Google Maps app failure on a Huawei Mate 9 Pro mobile phone running Android, detected using MRSimilar with Australian addresses. (a) Google Maps app returned "1 min" for walking from the current location to "Unit 9/890 Bourke Street, Zetland NSW." (b) Google Maps app returned "11 min" for driving from the current location to "Unit 8/890 Bourke Street, Zetland NSW." (c) Google Maps app returned "26 min" for walking from the current location to "Unit 8/890 Bourke Street, Zetland NSW." In (a), (b), and (c), the "current location" was basically the same. It was found that the location for the address "Unit 8/890 Bourke Street, Zetland NSW" generated by the Google Maps app in (b) and (c) was wrong: It was 2 km away from the actual location.

Fig. 4. Site visit to Unit 8/890 Bourke Street, Zetland, NSW, Australia, which confirmed that this was a valid physical address and that Unit 8 and the other units were located near each other.

We recognize that many so called "intelligent" systems (actually, their developers) might assume that they were smarter than the human users and therefore could automatically "correct" user input without even informing the users. This phenomenon was initially reported by Zhou et al. [14] where they studied web search engines and found that major search engines could automatically change user queries without explicitly informing the users, hence revealing a crucial deficiency in the software system's functional completeness.

4.2   Defects Detected by MRRestriction

Figure 5 shows a failure where Google Maps generated an infeasible route involving vehicular ferries. In Fig. 5(a), the user searched for a route and set the departing time to be "5:00 AM." Google Maps returned a route that will "arrive around 5:30 AM." The route involved the use of free vehicular ferry service operated by the government, which carries cars across the Clarence River. According to the government official website (Fig. 5(b)), the ferry operating hours start at 6:00 AM seven days a week. This means that the route shown in Fig. 5(a) is infeasible for the user's travel time (departing at 5:00 AM and arriving at the destination around 5:30 AM). Google Maps failed to identify this restriction and still recommended the user to use the (unavailable) ferry service. Actually, there was a nearby bridge that should be recommended instead for the given travel time.

This failure could be potentially due to incorrect ferry operating time data in the Google Maps database.

4.3   Defects Detected by MRSplit

When testing the Google Maps API against MRSplit, one of the source test cases involved an origin, a destination, and eight intermediate nodes. The corresponding follow-up test cases consisted of the nodes broken up in individual queries.

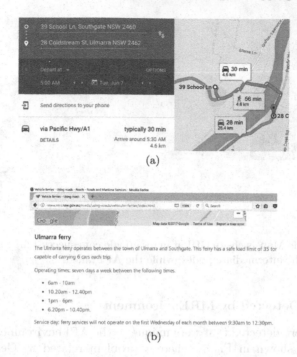

Fig. 5. Google Maps returned an infeasible route. (a) Excerpts of a screenshot showing that Google Maps returned a route involving vehicular ferry service for travel departing at "5:00 AM" and arriving "around 5:30 AM." (b) Ferry operating times: it starts at 6:00 AM, seven days a week.

For the source test case, Google Maps API returned UNKNOWN_ERROR as follows:

```
{
"routes"  :  [],
"status"  :  "UNKNOWN_ERROR"
}
```

In Google Maps API online documentation, an UNKNOWN_ERROR indicates that "a directions request could not be processed due to a server error" [35]. The "smaller" follow-up test cases, however, did not result in any error.

To further investigate this issue, we ran the test via the website GUI at maps.google.com, and the test passed without causing any failure or error, as shown in Fig. 6. This observation demonstrates that the Google Maps GUI and API are actually not the same, and in this test the API appeared to be more vulnerable to "large" input involving a large number of waypoint nodes. A further investigation shows that this failure could also be replicated using other "large" inputs.

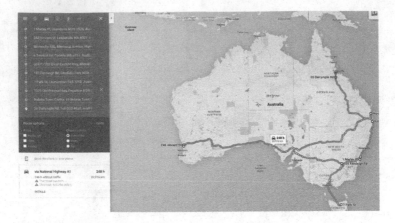

Fig. 6. The Google Maps website GUI at maps.google.com successfully passed this test case that had eight intermediate nodes, while the API failed.

4.4   Defects Detected by MREnvironment

More failures were detected that were unique to the API environment, an example of which is shown in Fig. 7, where a problem related to Geocoding was revealed. Geocoding is a process of converting addresses into geographic coordinates. Figure 7 shows that, for the test case under consideration, the API could not find the address, and therefore could not generate the geographic coordinates. As a result, the API could not generate a route, hence reporting a "NOT_FOUND" status.

When the same query was made via the Google Maps website GUI, it generated a route without causing any problem, as shown in Fig. 8. This observation again suggests that the Google Maps API was not as reliable as the website GUI.

```
"geocoded_waypoints" : [
{
"geocoder_status" : "ZERO_RESULTS"
},
{
"geocoder_status" : "OK",
"place_id" : "ChIJd0ShQqAZE2sRLCpaDBjrqUY",
"types" : [ "street_address" ]
}
],
"routes" : [],
"status" : "NOT_FOUND"
}
```

Fig. 7. Google Maps API failure: geocode not found.

Fig. 8. For the same test case, the Google Maps website GUI at maps.google.com has passed (as shown in this figure) whereas the Google Maps API failed (as shown in Fig. 7). This problem was detected using MREnvironment.

5 Discussions and Conclusion

In Sect. 1, we asked a research question: Can we have a practical and effective method of automatically testing mapping software in the face of the severe oracle problem?

To meet this challenge, we have proposed applying MT to test mapping systems, and have completed an initial case study using Google Maps. The results of this study provide an affirmative response to the research question. The detection of several types of real-life bugs in Google Maps further demonstrated the effectiveness of MT in testing "non-testable programs," i.e. programs that are difficult to test due to the lack of an oracle. Our testing approach can be used by developers for software verification, by users for software validation, and by independent testers for various quality assessment purposes.

Compared with the significance of the detection of major defects across several different environments (namely, the Google Maps mobile app, its web service API, and its GUI at maps.google.com), the testing cost we spent for Google Maps was relatively small. The reported failures could be caused by problems in the routing algorithms and/or the underlying databases. We have reported our findings to Google, who later confirmed that these issues were being addressed. In addition to Google Maps, we have also analysed the OpenStreetMap, and identified metamorphic relations that could be used to test such systems, from a developer's perspective. To implement these metamorphic relations to conduct testing will be an immediate research to be conducted in the near future.

This work suggests that mapping software can be considered as a special type of search engine, which accepts user queries and returns routes or driving directions. Previous results on search engine testing [14] can therefore be useful for the testing of mapping software. This research employed a useful general metamorphic relation (that is, a metamorphic relation pattern [32,34]) that is valid for both search engines and mapping software, namely, the software under test should return similar results for similar queries. This kind of pattern can be used to derive many concrete MRs. In future research, more effort should be made into the identification of MR patterns that can be used across different application domains.

This research contributes to the development of testing techniques for the navigation software of self-driving vehicles and autonomous robots and drones. Future research will be conducted at a larger scale by taking these systems into consideration.

Acknowledgements. This work was supported in part by a linkage grant of the Australian Research Council (project ID: LP160101691) and an Australian Government Research Training Program scholarship.

References

1. Brown, J., Zhou, Z.Q., Chow, Y.-W.: Metamorphic testing of navigation software: a pilot study with Google Maps. In: Proceedings of the 51st Annual Hawaii International Conference on System Sciences (HICSS-51), pp. 5687–5696 (2018). http://hdl.handle.net/10125/50602
2. The world factbook: Central Intelligence Agency (2013). https://www.cia.gov/library/publications/the-world-factbook/fields/2085.html
3. BuiltWith: Mapping usage statistics (2017). https://trends.builtwith.com/mapping
4. OpenStreetMap stats report: OpenStreetMap Foundation (2018). https://www.openstreetmap.org/stats/data_stats.html
5. Stats: OpenStreetMap Foundation (2018). https://wiki.openstreetmap.org/wiki/Stats
6. List of OSM-based services: OpenStreetMap Foundation (2018). https://wiki.openstreetmap.org/wiki/List_of_OSM-based_services
7. OSM Internet Links: OpenStreetMap Foundation (2018). https://wiki.openstreetmap.org/wiki/OSM_Internet_Links
8. Barr, E.T., Harman, M., McMinn, P., Shahbaz, M., Yoo, S.: The oracle problem in software testing: a survey. IEEE Trans. Softw. Eng. **41**(5), 507–525 (2015)
9. Wright, M., Stallings, D., Dunn, D.: The effectiveness of global positioning system electronic navigation. In: Proceedings of IEEE SoutheastCon, pp. 62–67 (2003)
10. Chen, T.Y., Tse, T.H., Zhou, Z.Q.: Fault-based testing without the need of oracles. Inf. Softw. Technol. **45**(1), 1–9 (2003)
11. Chen, T.Y., Kuo, F.-C., Zhou, Z.Q.: An effective testing method for end-user programmers. ACM SIGSOFT Softw. Eng. Notes **30**(4), 1–5 (2005)
12. Liu, H., Kuo, F.-C., Towey, D., Chen, T.Y.: How effectively does metamorphic testing alleviate the oracle problem? IEEE Trans. Softw. Eng. **40**(1), 4–22 (2014)
13. Lindvall, M., Ganesan, D., Árdal, R., Wiegand, R.E.: Metamorphic model-based testing applied on NASA DAT – an experience report. In: Proceedings of the 37th International Conference on Software Engineering (ICSE 2015), pp. 129–138 (2015)
14. Zhou, Z.Q., Xiang, S., Chen, T.Y.: Metamorphic testing for software quality assessment: a study of search engines. IEEE Trans. Softw. Eng. **42**(3), 264–284 (2016)
15. Segura, S., Fraser, G., Sanchez, A.B., Ruiz-Cortés, A.: A survey on metamorphic testing. IEEE Trans. Softw. Eng. **42**(9), 805–824 (2016)
16. Chen, T.Y., et al.: Metamorphic testing for cybersecurity. Computer **49**(6), 48–55 (2016)

17. Kanewala, U., Pullum, L.L., Segura, S., Towey, D., Zhou, Z.Q.: Message from the workshop chairs. In: Proceedings of the IEEE/ACM 1st International Workshop on Metamorphic Testing (ICSE MET 2016), in Conjunction with the 38th International Conference on Software Engineering (ICSE). ACM Press (2016)
18. Jarman, D.C., Zhou, Z.Q., Chen, T.Y.: Metamorphic testing for Adobe data analytics software. In: Proceedings of the IEEE/ACM 2nd International Workshop on Metamorphic Testing (ICSE MET 2017), in Conjunction with the 39th International Conference on Software Engineering (ICSE), pp. 21–27 (2017)
19. Ding, J., Hu, X.-H., Gudivada, V.: A machine learning based framework for verification and validation of massive scale image data. IEEE Trans. Big Data. https://doi.org/10.1109/TBDATA.2017.2680460
20. Pezzè, M., Young, M.: Software Testing and Analysis: Process, Principles, and Techniques. Wiley, New York (2008)
21. Chen, T.Y., Huang, D.H., Tse, T.H., Zhou, Z.Q.: Case studies on the selection of useful relations in metamorphic testing. In: Proceedings of the 4th Ibero-American Symposium on Software Engineering and Knowledge Engineering (JIISIC 2004). Polytechnic University of Madrid, pp. 569–583 (2004)
22. Goodman, J.L.: The space shuttle and GPS: a safety-critical navigation upgrade. In: Erdogmus, H., Weng, T. (eds.) ICCBSS 2003. LNCS, vol. 2580, pp. 92–100. Springer, Heidelberg (2003). https://doi.org/10.1007/3-540-36465-X_9
23. Nisner, P.D., Johannessen, R.: Ten million data points from TSO-approved aviation GPS receivers: results of analysis and applications to design and use in aviation. Navigation **47**(1), 43–50 (2000)
24. Elleuch, W., Wali, A., Alimi, A.M.: Mining road map from big database of GPS data. In: 14th International Conference on Hybrid Intelligent Systems (HIS), pp. 193–198. IEEE (2014)
25. Elleuch, W., Wali, A., Alimi, A.M.: Collection and exploration of GPS based vehicle traces database. In: 4th International Conference on Advanced Logistics and Transport (ICALT), pp. 275–280. IEEE (2015)
26. Chen, T.Y., Cheung, S.C., Yiu, S.M.: Metamorphic testing: a new approach for generating next test cases, Technical report HKUST-CS98-01, Department of Computer Science, Hong Kong Univ. of Science and Technology (1998)
27. Goodman, J.L.: A software perspective on GNSS receiver integration and operation. In: Rycroft, M. (ed.) Satellite Navigation Systems. Space Studies, vol. 8, pp. 119–126. Springer, Dordrecht (2003). https://doi.org/10.1007/978-94-017-0401-4_13
28. Luxen, D., Vetter, C.: Real-time routing with OpenStreetMap data. In: Proceedings of the 19th ACM SIGSPATIAL International Conference on Advances in Geographic Information Systems (GIS 2011), pp. 513–516. ACM (2011)
29. Graf, F., Kriegel, H.-P., Renz, M., Schubert, M.: MARiO: multi-attribute routing in open street map. In: Pfoser, D. (ed.) SSTD 2011. LNCS, vol. 6849, pp. 486–490. Springer, Heidelberg (2011). https://doi.org/10.1007/978-3-642-22922-0_36
30. Cipeluch, B., Jacob, R., Winstanley, A., Mooney, P.: Comparison of the accuracy of OpenStreetMap for Ireland with Google Maps and Bing Maps. In: Proceedings of the 9th International Symposium on Spatial Accuracy Assessment in Natural Resources and Environmental Sciences, pp. 337–341 (2010)
31. Chen, T.Y., et al.: Metamorphic testing: a review of challenges and opportunities. ACM Comput. Surv. **51**(1), 4:1–4:27 (2018)
32. Zhou, Z.Q., Sun, L., Chen, T.Y., Towey, D.: Metamorphic relations for enhancing system understanding and use. IEEE Trans. Softw. Eng. https://doi.org/10.1109/TSE.2018.2876433

33. Zhou, Z.Q., Sun, L.: Metamorphic testing of driverless cars. Commun. ACM **62**(3), 61–67 (2019)
34. Segura, S., Parejo, J.A., Troya, J., Ruiz-Cortés, A.: Metamorphic testing of REST-ful Web APIs. IEEE Trans. Softw. Eng. **44**(11), 1083–1099 (2018)
35. Google Maps Directions API: Google (2016). https://developers.google.com/maps/documentation/directions/intro

Multi-criteria Recommendations by Using Criteria Preferences as Contexts

Yong Zheng[✉]

School of Applied Technology, Illinois Institute of Technology,
Chicago, IL 60616, USA
yong.zheng@iit.edu

Abstract. Recommender system is a well-known information system which assists decision making by producing recommendations tailored to user preferences. Multi-criteria recommender systems (MCRS) additionally take user preferences in multiple criteria into account, in order to better generate recommendations. The major challenge in MCRS is the process of aggregating user ratings in the multiple criteria. We claim that user preferences in these criteria can be considered as contexts, so that the overall taste on an item can be estimated by a process of context-aware predictions. In this paper, we exploit and summarize different methods which produce the recommendations by using criteria preferences as context information. We examine these methods based on three real-world data sets. Our experimental results demonstrate the effectiveness of these algorithms in the rating prediction task, in comparison with the state-of-the-art multi-criteria recommendation approaches.

Keywords: Recommender system · Multi-criteria · Context · Context-aware

1 Introduction and Motivations

Recommender systems is an effective solution to alleviate the problem of information overload and assist decision making. A traditional recommender may produce a list of recommendations tailored by user preferences. It has been widely applied to online streaming (e.g., Netflix, Spotify) [7,17], E-commerce (e.g., Amazon.com) [18,20], social networks (e.g., Facebook) [5,12], tourism (e.g., TripAdvisor) [9,29], educations [5,14,26], etc.

Several novel recommender systems were proposed to improve the recommendations and adapt to new applications. One of them is the context-aware recommender systems (CARS) [3] which leverage the value of recommendations by exploiting context information (e.g., time, location, weather, etc) that affects user preferences. Context-awareness is necessary in the area of recommender systems, since a user's taste may vary from contexts to contexts. For example, a user may choose a different type of the movie if he or she is going to watch the movie

T. A. Majchrzak et al. (Eds.): *Towards Integrated Web, Mobile,*
and IoT Technology, LNBIP 347, pp. 21–35, 2019.
https://doi.org/10.1007/978-3-030-28430-5_2

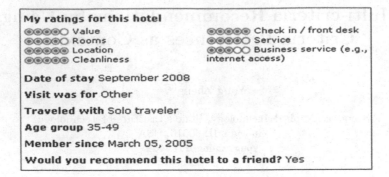

Fig. 1. Multi-criteria ratings on TripAdvisor.com

with *kids* rather than with *partner*. Or, a user may perform a different outdoor activity when it is *at weekend* instead of *the weekday*. *Companion* and *the day of the week* are two context variables in these two examples. CARS additionally take these context information into consideration so that the recommendations can be adapted to these context situations.

Another interesting one is the multi-criteria recommender systems (MCRS) [2] which take advantage of user preferences in multiple criteria. Take the case of TripAdvisor.com as shown by Fig. 1 for example, a user can give ratings on multiple criteria (e.g., room size, cleanness, customer service, etc) in addition to the overall rating on a hotel. MCRS try to aggregate these multi-criteria ratings in order to better predict a user's overall taste on the item. The practice of MCRS has been successfully applied to TripAdvisor [9], Yahoo!Movie [9,22], OpenTable [6], and so forth.

The major challenge in MCRS is the process of aggregating user preferences in multiple criteria. Recently, we propose that user preferences in these criteria can be viewed as contexts [22,23] so that context-awareness could be involved in the multi-criteria recommendations. Different methods which implement this idea have been proposed. This paper extends our previous work [23] presented at the track on Web Technologies, ACM symposium on Applied Computing in 2017[1]. More specifically, we exploit and summarize these approaches, and perform empirical evaluations on three real-world data sets in comparison with the state-of-the-art multi-criteria recommendation algorithms. Out experimental results demonstrate the effectiveness of the multi-criteria recommendation models using criteria preferences as contexts.

2   Related Work

In this section, we introduce the related work in CARS and MCRS. In addition, we describe the aggregation-based multi-criteria recommendation approach,

---

[1] ACM SAC 2017, https://www.sigapp.org/sac/sac2017/.

upon which we discuss the extensions by using criteria preferences as contexts in the following sections.

2.1 Context-Aware Recommender Systems

The standard formulation of the collaborative recommendation problem begins with a two-dimensional matrix of ratings, organized by user and item: *Users × Items → Ratings*. Recommendation then becomes a prediction problem, interpolating new ratings not present in the original matrix. Context-aware recommender systems add contextual variables to this question, making use of the context in which a rating was made and the context in which a recommendation is sought, in order to add nuance to the resulting recommendations [3]. The updated rating prediction in CARS becomes *Users × Items × Contexts → Ratings*.

Table 1. Contextual ratings on movies

User	Item	Rating	Time	Location	Companion
U_1	T_1	3	weekend	home	alone
U_1	T_1	5	weekend	cinema	partner
U_1	T_1	?	weekday	home	family

The "context" information usually refers to the dynamic variables which may change when a same activity was performed repeatedly [21]. For example, the time, location, as well as a user's emotions may vary every time when he or she is going to watch movies or listen to songs. Take the context-aware movie ratings in Table 1 for example, we introduce the terminologies in CARS as follows. There is one user U_1, one movie T_1, and three context variables – Time (weekend or weekday), Location (at home or cinema) and Companion (alone, partner, family). In the following discussion, we use *context dimension* to denote the contextual variable, e.g. "Location". The term *context condition* refers to a specific value in a dimension, e.g. "home" and "cinema" are two contextual conditions in "Location". The *contexts* or *context situation* is, therefore, a set of contextual conditions, e.g. {*weekend, home, family*}.

There are three ways to build a contextual recommender. *Contextual pre-filtering*, such as the splitting-based methods [30], will use the context information to filter out irrelevant rating profiles and apply the traditional recommendation algorithms to produce the recommendations. By contrast, *contextual post-filtering* methods produce recommendations without considering contexts, and then utilize contexts to adjust the predicted ratings or re-rank the items [15,24]. *Contextual modeling* [4,28,31] is the most complicated strategy, while contexts are directly incorporated into the predictive models. There is limited research on post-filtering, while pre-filtering and contextual modeling methods are usually the more popular and effective ways to build CARS.

2.2   Multi-criteria Recommender Systems

MCRS tries to additionally take multi-criteria ratings into account to build better recommenders. An example of data in MCRS can be shown by Table 2. The *rating* refers to the users' overall rating on the items. We also have users' ratings on multiple criteria, such as room, check-in and service. These ratings are referred as *criteria preferences*. Given a user U_3 and an item T_1 shown in Table 2, the rating prediction task in MCRS is how to predict U_3's overall rating on T_1. Note that we also do not know U_3's multi-criteria ratings on T_1.

Table 2. Example of rating data from TripAdvisor

User	Item	Rating	Room	Check-in	Service
U_1	T_1	3	3	4	3
U_2	T_2	4	4	4	5
U_3	T_1	?	?	?	?

The *heuristic* and *model-based* methods have been developed to produce multi-criteria recommendations. In the heuristic approaches, the multi-criteria ratings can be used to better calculate user-user or item-item similarities in the collaborative filtering algorithms. For example, Adomavicius et al. [1] proposed the average and the worst-case as two strategies to aggregate the traditional user-user similarities and the similarities obtained from criteria preferences. Another contribution on improving user-user similarities was made by Manouselis and Costopoulou [13], in which the similarities between users are obtained using multi-criteria ratings, and the rest of the recommendation process can be the same as in single-criterion rating systems.

By contrast, the model-based approaches may build a predictive model to estimate a user' overall rating on the item from the observed multi-criteria ratings. Adomavicius et al. [1] proposed a linear-aggregation based multi-criteria recommendation method, in which the overall rating can be estimated by a linear regression model using the predicted criteria preferences as independent variables. Sahoo et al. [16] proposed a probabilistic model by extending the flexible mixture model (FMM) [19] to MCRS. Particularly, the dependency among multiple criteria can be incorporated into the FMM structure and further improve the recommendation performance. Most recently, we propose the utility-based multi-criteria recommendation algorithm [27] for top-N recommendations. More specifically, we propose to use a utility score to rank the candidate items, while the score can be computed as the similarity between user expectations and criteria preferences. We employ the listwise ranking to learn user expectations by maximizing the normalized discounted cumulative gain in the top-N recommendations.

2.3 Preliminary: The Aggregation-Based Recommendation Approach

The aggregation-based multi-criteria recommendation approach [1,10] is one of the most popular and standard ways to produce multi-criteria recommendations. In this section, we introduce the general steps in these aggregation-based approaches, and further discuss the models using criteria preferences as contexts which were built upon these aggregation-based methods in Sect. 3.

$$R_0 = f(R_1, R_2, ..., R_k) \tag{1}$$

Generally, the aggregation-based approach builds an aggregation function f that represents the relationship between the overall rating R_0 and multi-criteria ratings (e.g., $R_1, R_2, ..., R_k$), as shown in Eq. 1. Accordingly, there are two steps to perform these aggregation-based approaches:

1. **Multi-Criteria Rating Predictions:** First of all, we need to predict the rating on each individual criterion. As shown by the Table 2, we need to predict how U_3 will rate the item T_1 in the criteria *room, check-in* and *service* respectively. The predicted criteria preferences can be used for aggregations in the next step.
2. **Rating Aggregations:** Once the ratings on each criterion (e.g., R_1, $R_2, ..., R_k$) have been predicted, we can figure out the aggregation function f to utilize these predicted criteria preferences to estimate the overall rating on the items.

$$R_0 = w_1 * R_1 + w_2 * R_2 + ... + w_k * R_k + t \tag{2}$$

The simplest way for the aggregation is a linear regression [1] which assumes there is a linear relationship between the multi-criteria ratings and the overall rating. R_0 can be estimated by a multiple linear regression as shown by Eq. 2, where we assign a weight (e.g., w_k) to each criterion, and finally learn these weights, as well as the intercept t by minimizing the sum of the squared prediction errors.

3 Methodologies by Using Criteria Preferences as Contexts

In this section, we exploit and discuss different ways to build the multi-criteria recommendation models by using criteria preferences as contexts.

First of all, we discuss the reason why criteria preferences can be viewed as contexts. Take hotel booking on TripAdvisor.com as shown by Fig. 2 for example, we assume a user is viewing the hotel reviews in order to learn whether the hotel can meet his or her expectations. The user may already have a judgement against the hotel in different criteria by referring to other users' perspectives. In the review above, the user may learn the location of the hotel is very convenient, but

Fig. 2. Example of hotel reviews on TripAdvisor.com

the room is not clean enough. In terms of the psychology, the user may already "rate" the hotel on different criteria in their mind, such as room cleanliness - 2 star (i.e., very bad), location - 4 star (i.e., good), and so forth. Afterwards the user will make a final decision about whether they will reserve rooms in this hotel, according to their judgements after reviews reading. In this case, the users' judgements on different aspects of the hotel (i.e., room cleanliness and location) can be viewed as the contextual situation in which user will make a final decision. It transforms the process of predicting the overall rating to a context-aware procedure. Namely, the overall rating predictions can be formulated to such a problem: given the contextual situation – how much a user likes different aspects of the items, the system will predict how the user will like or dislike the item.

As mentioned in the previous section, there are two steps in the aggregation-based multi-criteria recommendations. As a result, we can utilize criteria preferences as contexts in each of these two steps respectively. More specifically, we discuss these methodologies as follows.

3.1   Multi-criteria Rating Predictions Using Criteria Preferences as Contexts

In fact, we can predict the multi-criteria ratings independently or dependently. In the *independent* method, we predict the rating on each criterion by using the rating matrix associated with each criterion. Take Table 2 for example, to predict how a user will rate an item in the criterion "Room", we use the rating matrix <User, Item, Room> only. This method is simple and straightforward, but it ignores the correlations among the multiple criteria.

However, there could be correlations or dependencies among different criteria. For example, a hotel close to train station may be far away from the city center. It could also be noise in the hotel rooms due to its locations. As a result, the rating in "location" and "quietness" may be correlated. In multi-criteria recommendations, both FMM [19] and criteria chains [22] can incorporate the dependency among criteria into the recommendation models. We particularly discuss criteria chains in this section, since it is easier to be integrated in the aggregation-based multi-criteria recommendations.

The major advantage of criteria chains is that it additionally considers the correlations among multiple criteria. First of all, it defines the sequence of the criteria as a chain by using information gain as the impurity criterion. We assume the sequence is "Service - Room - Check-in" in the data shown by Table 2. In this case, the variable "Service" is the dimension with largest impurity measured by information gain, and "Check-in" is the one with least impurity. Once the chain is defined, we will predict the rating in the dimension "Service" first. The predicted rating in "Service" will be reviewed as context to be used to predict the rating in the next dimension in the chain which is "Room". Finally, the predicted ratings in "Service" and "Room" will be considered as contexts to predict the rating in "Check-in". In other words, only the rating prediction on the first criterion in the chain is a context-free process. The subsequent rating estimations on other criteria are context-aware rating predictions, while any rating prediction functions in the context-aware recommenders can be used for this purpose. Once the multi-criteria ratings have been predicted, they can be utilized in the process of rating aggregations to estimate the user's taste on the items. Note that the predicted multi-criteria ratings are numerical values with decimals. To reduce the sparsity problem in context conditions, we can either cast or round these values to integers in order to improve the performance of context-aware predictions. In our experiments, rounding is the best option for this operation.

3.2   Rating Aggregations Using Criteria Preferences as Contexts

Alternatively, we can also use criteria preferences as contexts in the step of rating aggregations. As mentioned previously, the predicted criteria preferences will be used to aggregate and estimate a user's overall rating on an item. Linear regression has been applied as one of these aggregating functions. By contrast, we propose the *context aggregations* [23]. More specifically, we view all of the predicted criteria preferences as context information, and adopt a context-aware recommender to predict the overall rating. In this case, we can predict the multi-criteria ratings independently or dependently, and we round the predicted ratings to integers which will be considered as context conditions. A context-aware recommender will estimate the overall rating by using these predicted criteria preferences as contexts.

3.3   Summary

According to the descriptions above, we have two options in the step of multi-criteria rating predictions – independent and dependent methods. We also have two options in the step of rating aggregations – linear and context aggregations. It enables us to build four different multi-criteria recommendation models:

– **Independent Linear Aggregation (ILA):** We predict the multi-criteria ratings independently, and then use linear aggregation to estimate the overall rating. Note that we did not view criteria preferences as contexts in this method. ILA is a standard aggregation-based approach and baseline method to produce multi-criteria recommendations.

- **Independent Context Aggregation (ICA):** In this method, we predict the multi-criteria ratings independently, and use the context aggregation to estimate the overall rating. ICA is equivalent to the criteria-independent contextual models described in [22]. It is also the same as the full contextual model in [23].
- **Dependent Linear Aggregation (DLA):** Or, we can predict the multi-criteria ratings dependently by using the idea of criteria chains, but we finally use linear aggregation to estimate the overall rating. DLA is equivalent to the linear aggregation model using criteria chains described in [22].
- **Dependent Context Aggregation (DCA):** Finally, we can predict the multi-criteria ratings dependently and use context aggregation to estimate the overall rating. In this case, the criteria preferences are viewed as contexts in all of the two steps. DCA is equivalent to the contextual model using criteria chains described in [22].

3.4   Traditional and Context-Aware Rating Predictions

Note that we need a traditional and a context-aware rating prediction function to estimate the multi-criteria ratings independently and dependently. In this section, we briefly introduce these predictive methods used in our paper.

We utilize the biased matrix factorization (MF) [11] as the predictive function for the independent predictions. MF is a latent-factor based learning technique and it is usually considered as a standard and popular benchmark method in the area of recommender systems. The rating prediction function is shown by Eq. 3.

$$\hat{r}_{ui} = \mu + b_u + b_i + p_u^T q_i \tag{3}$$

μ refers to the global average rating, while b_u and b_i are the bias associated with user u and item i respectively. p_u and q_i are the latent-factor vector which can represent u and i respectively. The MF will learn these parameters by minimizing sum of squared errors by using stochastic gradient descent (SGD) as the optimizer. The L_2 norms are usually added into the loss function as the regularization terms in order to alleviate overfitting. The loss function can be described by Eq. 4, where u, i is an entry in the training set T, and λ is the regularization rate. r_{ui} and $\hat{r}_{ui}$ are the real rating and predicted rating for the entry u, i respectively. We use the implementation of MF algorithm in LibRec [8] which is an open-source library for traditional recommendations.

$$\underset{p*,q*,b*}{Minimize} \sum_{(u,i) \in T} (r_{ui} - \hat{r}_{ui})^2 + \lambda(||p_u||^2 + ||q_i||^2 + b_u^2 + b_i^2) \tag{4}$$

In our work, we select context-aware matrix factorization (CAMF) [4] as the model to perform dependent predictions. It is because CAMF is an effective and standard contextual modeling technique which learns rating deviations in the process of matrix factorization. There are different versions of CAMF, while we use the basic one which learns the rating deviation in each context condition individually. The rating prediction in the CAMF can be shown as Eq. 5.

$$\hat{r}_{uic_{k,1}c_{k,2}...c_{k,L}} = \mu + b_u + b_i + \sum_{j=1}^{L} B_{c_{k,j}} + \vec{p_u} \cdot \vec{q_i} \tag{5}$$

$\hat{r}_{uic_{k,1}c_{k,2}...c_{k,L}}$ denotes the predicted rating in a specific context situation c_k. Assume there are L contextual dimensions in total, $c_k = \{c_{k,1}c_{k,2}...c_{k,L}\}$ is used to describe the contextual situation, where $c_{k,j}$ denotes the contextual condition in the jth context dimension. Therefore, $B_{ijc_{k,j}}$ indicates the contextual rating deviation associated with item i and the contextual condition in the jth dimension. $\vec{p_u}$ and $\vec{q_i}$ represent the user vector and item vector for user u and item i respectively. They are the standard components in the technique of matrix factorization. μ is the global average rating in the data, while b_u and b_i represent the user bias and item bias. $B_{c_{k,j}}$ denote the bias or rating deviation in the context condition $c_{k,j}$. We adopt the SGD as the optimizer to learn these parameters. We use the implementation of the CAMF algorithm in the CARSKit [32] which is an open-source library for context-aware recommendations.

In addition, we need a context-aware recommender for the purpose of context aggregations. The CAMF approach described above is also adopted in the process of context aggregations.

4   Experiments and Results

4.1   Data Sets

There are not many data sets with multi-criteria ratings for public research, and we use the following three real-world data sets for the purpose of empirical evaluations:

– **TripAdvisor data:** This data was crawled by Jannach et al. [9]. The data was collected through a Web crawling process which collects users' ratings on hotels located in 14 global metropolitan destinations, such as London, New York, Singapore, etc. There are 22,130 ratings given by 1,502 users and 14,300 hotels. Each user gave at least 10 ratings which are associated with multi-criteria ratings on seven criteria: *value* for the money, quality of *rooms*, convenience of the hotel *location*, *cleanliness* of the hotel, experience of *check-in*, overall quality of *service* and particular *business services*.
– **Yahoo!Movie data:** This data was obtained from Yahoo!Movies by Jannach et al. [9]. There are 62,739 ratings given by 2,162 users on 3,078 movies. Each user left at least 10 ratings which are associated with multi-criteria ratings on four criteria: *direction, story, acting* and *visual effects*.

- **ITMLearning data:** This is a data set in the area of educational domain which was collected by the user surveys [25]. There are 3,306 ratings given by 269 students on 70 projects. In addition to the overall ratings, each student will rate the selected projects in three criteria (i.e., App, Data and Ease): how interesting the application area is (App), how convenient the data processing will be (Data), how easy the whole project is (Ease) by using this data set.

Both of the overall ratings and the multi-criteria ratings are in the scale 1 to 5 in the three data sets mentioned above.

4.2 Baselines and Evaluations

We compare the approaches which utilize criteria preferences as contexts (i.e., ICA, DLA, DCA) with other multi-criteria recommendation models[2] which can be listed as follows:

- The model *MF* is the biased matrix factorization model [11] by using the rating matrix <User, Item, Ratings> only without considering multi-criteria ratings.
- The model *ILA* [1] is the approach we introduced in Sect. 3.3. It is the standard aggregation-based multi-criteria recommendation method which predicts the multi-criteria ratings independently and uses a linear aggregation to estimate a user's overall rating on an item.
- The model *FMM* [16] is a probabilistic recommendation approach based on the flexible mixture model (FMM) [19]. It is another effective approach which additional takes the correlations among the criteria into considerations.

We use a 5-fold cross validation for each data and evaluate the recommendation performance in the rating prediction task by using mean absolute error (MAE) as the evaluation metric. MAE is a popular metric for rating predictions, and it can be described in Eq. 6, where T_e represents the testing set, $R_{u,i}$ is the real rating given by user u on item i, and $\widehat{R}_{u,i}$ is our predicted rating in the same setting. From the perspective of the algorithms, we'd like to build or learn a predictive recommendation model based on the training set in order to minimize the MAE in the testing set. We try different parameters in each recommender to find the best setting and report the optimal results in MAE.

$$MAE = \frac{1}{|T_e|} \sum_{(u,i)\epsilon T_e} |\widehat{R}_{u,i} - R_{u,i}| \tag{6}$$

4.3 Experimental Results and Findings

By predicting multi-criteria ratings dependently, we need to identify the chain of the criteria in the DLA and DCA approaches. Our previous work [22] has

---

[2] We evaluate these models based on the rating prediction task in this paper. The utility-based multi-criteria recommendation model [27] was not selected as one of the baseline methods since it can only be used for top-N recommendations.

demonstrated that information gain was the best metric to rank the criteria. Based on the information gain, the criteria chain for each data can be described as follows (Table 3):

Table 3. Sequence of criteria chains

Data	Sequence of criteria in criteria chains
TripAdvisor	Value - Rooms - Service - Cleanliness - Check-in - Location - Business Services
Yahoo!Movie	Direction - Story - Acting - Visuals
ITMLearning	App - Data - Ease

The experimental results based on the TripAdvisor and Yahoo!Movie data sets can be depicted by Fig. 3. In addition to the MAE results, we use paired two-sample hypothesis testing to examine to examine the significance of the results between the approaches which utilize criteria preferences as contexts (i.e., ICA, DLA and DCA) and other baseline methods (i.e., MF, ILA and FMM). More specifically, we use "*" to indicate the significant advantage of the selected approaches (i.e., ICA, DLA and DCA) and the standard baseline methods including MF and ILA. In addition, we mark the approach with an additional "*" if the approach is able to significantly outperform the FMM method which is the most of effective multi-criteria recommendation approaches.

We are able to observe that most of the approaches which utilize criteria preferences as contexts can outperform the standard baselines (i.e., MF and ILA), except the ICA method in the Yahoo!Movie data set. ICA produces lower MAE than the ILA approach, but it fails the significance test. Furthermore, DCA is the only approach which can beat FMM significantly. And DCA becomes the best performing algorithm among all of these multi-criteria recommendation models. These results demonstrate that it is effective to utilize criteria preferences as contexts in the multi-criteria recommendations.

Figure 4 presents the results on the ITMLearning data set. The results are significantly different from the ones in the TripAdvisor and Yahoo!Movie data. First of all, ICA and FMM are the best performing algorithms. DLA and DCA work worse than ICA, and they even did not produce better results than the standard methods (i.e., MF and ILA). ICA outperforms the standard baselines, which confirms the advantage of the methods that utilize criteria preferences as contexts.

We further investigate the ITMLearning data to understand why DLA and DCA fail to produce better results. Recall that we have three criteria in this data – App, Data and Ease. However, it is not necessary to say a student may like one project if they rate higher in these three criteria. We found that some students preferred to select an easy project, while some others would like to choose more challenging projects. It results in conflicting rating patterns. For example, a student may give a higher overall rating to a project. It is possible

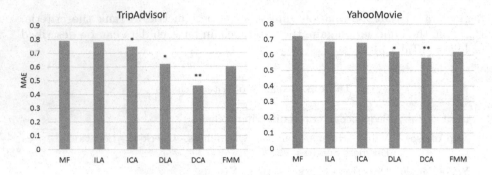

Fig. 3. Experimental results based on MAE

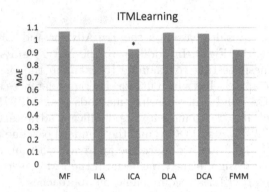

Fig. 4. Experimental results of the ITMLearning data

that he gave higher and positive ratings to "App" and "Data" but lower rating in "Ease", since he prefers to select more challenging projects rather than the easy ones. Therefore, the predicted multi-criteria ratings by the dependent method (i.e., criteria chains which is used in DLA and DCA) may be not that accurate due to these conflicting interests in the criteria preferences. By contrast, in other applications, such as TripAdvisor and Yahoo!Movie, a user may gave a higher overall rating, if his or her ratings on multiple criteria are all positive and higher.

Figure 5 shows the MAE results of the step of multi-criteria rating predictions. We can observe that the MAE results are generally reduced if we use the dependent methods. Note that the MAE results for the "Value" in the TripAdvisor and the "Direction" in the Yahoo!Movie data are the same by the independent and dependent methods, since these criteria are the first one according to the sequence in the criteria chain. By contrast, Fig. 6 presents the results on the ITMLearning data, where we can observe that the MAE values are increased by using the dependent method. It explains the reason why DLA and DCA did not perform well in this data.

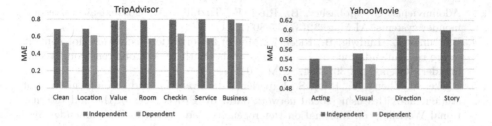

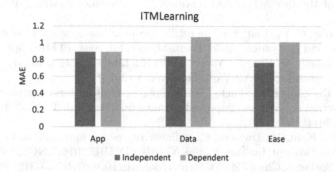

Fig. 5. MAE of predicted criteria preferences

Fig. 6. MAE of predicted criteria preferences on the ITMLearning data

5   Conclusions and Future Work

In this paper, we exploit and summarize different ways to build multi-criteria recommendation models by using the criteria preferences as context information. Our experimental results demonstrate the effectiveness of these approaches (i.e., ICA, DLA, DCA) in comparison with the state-of-the-art multi-criteria recommendation algorithms. The approaches which utilize the dependent method to predict the multi-criteria ratings failed to outperform the baseline methods in the ITMLearning data, due to the special characteristics of the multiple criteria and conflicting interests in the data. We will seek solutions to address these issues in our future work.

Acknowledgement. We thank Master student Shephalika Shekhar in Illinois Institute of Technology for her assistance with parts of experimental results on the ITMLearning data set.

References

1. Adomavicius, G., Kwon, Y.: New recommendation techniques for multicriteria rating systems. IEEE Intell. Syst. **22**(3), 48–55 (2007)
2. Adomavicius, G., Kwon, Y.: Multi-criteria recommender systems. In: Ricci, F., Rokach, L., Shapira, B. (eds.) Recommender Systems Handbook, pp. 847–880. Springer, Boston (2015). https://doi.org/10.1007/978-1-4899-7637-6_25

3. Adomavicius, G., Mobasher, B., Ricci, F., Tuzhilin, A.: Context-aware recommender systems. AI Mag. **32**(3), 67–80 (2011)
4. Baltrunas, L., Ludwig, B., Ricci, F.: Matrix factorization techniques for context aware recommendation. In: Proceedings of the Fifth ACM Conference on Recommender Systems, pp. 301–304. ACM (2011)
5. Burke, R., Zheng, Y., Riley, S.: Experience discovery: hybrid recommendation of student activities using social network data. In: Proceedings of the 2nd International Workshop on Information Heterogeneity and Fusion in Recommender Systems, pp. 49–52. ACM (2011)
6. Das, S.: Making meaningful restaurant recommendations at OpenTable. In: Proceedings of the 9th ACM Conference on Recommender Systems, p. 235. ACM (2015)
7. Gomez-Uribe, C.A., Hunt, N.: The netflix recommender system: algorithms, business value, and innovation. ACM Trans. Manag. Inf. Syst. (TMIS) **6**(4), 13 (2016)
8. Guo, G., Zhang, J., Sun, Z., Yorke-Smith, N.: LibRec: a Java library for recommender systems. In: UMAP Workshops, vol. 4 (2015)
9. Jannach, D., Zanker, M., Fuchs, M.: Leveraging multi-criteria customer feedback for satisfaction analysis and improved recommendations. Inf. Technol. Tour. **14**(2), 119–149 (2014)
10. Jhalani, T., Kant, V., Dwivedi, P.: A linear regression approach to multi-criteria recommender system. In: Tan, Y., Shi, Y. (eds.) DMBD 2016. LNCS, vol. 9714, pp. 235–243. Springer, Cham (2016). https://doi.org/10.1007/978-3-319-40973-3_23
11. Koren, Y., Bell, R., Volinsky, C.: Matrix factorization techniques for recommender systems. Computer **42**(8), 30–37 (2009)
12. Ma, H., Zhou, D., Liu, C., Lyu, M.R., King, I.: Recommender systems with social regularization. In: Proceedings of the Fourth ACM International Conference on Web Search and Data Mining, pp. 287–296. ACM (2011)
13. Manouselis, N., Costopoulou, C.: Experimental analysis of design choices in multiattribute utility collaborative filtering. Int. J. Pattern Recognit. Artif. Intell. **21**(02), 311–331 (2007)
14. Manouselis, N., Drachsler, H., Verbert, K., Santos, O.C.: Recommender Systems for Technology Enhanced Learning: Research Trends and Applications. Springer, New York (2014). https://doi.org/10.1007/978-1-4939-0530-0
15. Panniello, U., Tuzhilin, A., Gorgoglione, M., Palmisano, C., Pedone, A.: Experimental comparison of pre-vs. post-filtering approaches in context-aware recommender systems. In: Proceedings of the Third ACM Conference on Recommender Systems, pp. 265–268. ACM (2009)
16. Sahoo, N., Krishnan, R., Duncan, G., Callan, J.: Research note-the Halo effect in multicomponent ratings and its implications for recommender systems: the case of yahoo! movies. Inf. Syst. Res. **23**(1), 231–246 (2012)
17. Sánchez-Moreno, D., Zheng, Y., Moreno-García, M.N.: Incorporating time dynamics and implicit feedback into music recommender systems. In: 2018 IEEE/WIC/ACM International Conference on Web Intelligence (WI), pp. 580–585. IEEE (2018)
18. Schafer, J.B., Konstan, J., Riedl, J.: Recommender systems in e-commerce. In: Proceedings of the 1st ACM Conference on Electronic Commerce, pp. 158–166. ACM (1999)
19. Si, L., Jin, R.: Flexible mixture model for collaborative filtering. In: Proceedings of the 20th International Conference on Machine Learning (ICML 2003), pp. 704–711 (2003)

20. Smith, B., Linden, G.: Two decades of recommender systems at amazon.com. IEEE Internet Comput. **21**(3), 12–18 (2017)
21. Zheng, Y.: A revisit to the identification of contexts in recommender systems. In: Proceedings of the Conference on Intelligent User Interfaces Companion, pp. 133–136. ACM (2015)
22. Zheng, Y.: Criteria chains: a novel multi-criteria recommendation approach. In: Proceedings of the 22nd ACM International Conference on Intelligent User Interfaces, pp. 29–33. ACM (2017)
23. Zheng, Y.: Situation-aware multi-criteria recommender system: using criteria preferences as contexts. In: Proceedings of the ACM Symposium on Applied Computing, pp. 689–692. ACM (2017)
24. Zheng, Y.: Context-aware mobile recommendation by a novel post-filtering approach. In: The Thirty-First International Flairs Conference (2018)
25. Zheng, Y.: Personality-aware decision making in educational learning. In: Proceedings of the 23rd ACM International Conference on Intelligent User Interfaces Companion, p. 58. ACM (2018)
26. Zheng, Y.: Multi-stakeholder personalized learning with preference corrections. In: Proceedings of 18th IEEE International Conference on Advanced Learning Technologies (ICALT) (2019)
27. Zheng, Y.: Utility-based multi-criteria recommender systems. In: Proceedings of the 34th Annual ACM Symposium on Applied Computing. ACM (2019)
28. Zheng, Y., Anna Jose, A.: Context-aware recommendations via sequential predictions. In: Proceedings of the 34th Annual ACM Symposium on Applied Computing. ACM (2019)
29. Zheng, Y., Burke, R., Mobasher, B.: Differential context relaxation for context-aware travel recommendation. In: Huemer, C., Lops, P. (eds.) EC-Web 2012. LNBIP, vol. 123, pp. 88–99. Springer, Heidelberg (2012). https://doi.org/10.1007/978-3-642-32273-0_8
30. Zheng, Y., Burke, R., Mobasher, B.: Splitting approaches for context-aware recommendation: an empirical study. In: Proceedings of the 29th Annual ACM Symposium on Applied Computing, pp. 274–279. ACM (2014)
31. Zheng, Y., Mobasher, B., Burke, R.: CSLIM: contextual SLIM recommendation algorithms. In: Proceedings of the ACM Conference on Recommender Systems, pp. 301–304. ACM (2014)
32. Zheng, Y., Mobasher, B., Burke, R.: CARSKit: a Java-based context-aware recommendation engine. In: 2015 IEEE International Conference on Data Mining Workshop, pp. 1668–1671. IEEE (2015)

Towards Pluri-Platform Development: Evaluating a Graphical Model-Driven Approach to App Development Across Device Classes

Christoph Rieger[✉] and Herbert Kuchen

ERCIS, University of Münster, Münster, Germany
{christoph.rieger,kuchen}@uni-muenster.de

Abstract. The domain of mobile apps encompasses a fast-changing ecosystem of platforms and vendors in which new classes of heterogeneous app-enabled devices are emerging. To digitize everyday work routines, business apps are used by many non-technical users. However, designing apps is mostly done according to traditional software development practices, and further complicated by the variability of device capabilities. To empower non-technical users to participate in the creation of supportive apps, graphical domain-specific languages can be used. Consequently, we propose the Münster App Modeling Language (MAML) to specify business apps through graphical building blocks on a high level of abstraction. In contrast to existing process modelling notations, these models can directly be transformed into apps for multiple platforms across different device classes through code generators without the need for manual programming. To evaluate the comprehensibility and usability of MAML's DSL, two studies were performed with software developers, process modellers, and domain experts.

Keywords: Graphical domain-specific language ·
Model-driven software development · Business app · Cross-platform

1 Introduction

The opportunities of Model-Driven Software Development (MDSD) with regard to increased efficiency and flexibility have been studied extensively in the past years. The use of MDSD techniques is one approach to counteract the variety of platforms, programming languages, and human-interface guidelines found in the domain of mobile business apps. Several approaches have been researched in academic literature, including MD^2 [34], Mobl [26], and AXIOM [30]. Those approaches provide cross-platform development functionalities with one common model for multiple target platforms using textual domain-specific languages (DSLs) to specify apps. Whereas these approaches significantly ease the development of apps and thus also support current trends such as "Bring your own

© Springer Nature Switzerland AG 2019
T. A. Majchrzak et al. (Eds.): *Towards Integrated Web, Mobile,
and IoT Technology*, LNBIP 347, pp. 36–66, 2019.
https://doi.org/10.1007/978-3-030-28430-5_3

device" [58], the actual creation of apps is still restricted to users with programming skills [37]. Business apps focus on specific tasks to be accomplished by employees. Therefore, a centralized definition of such processes aligns well with traditional software development practices but may deviate from the end user's needs. Consequently, the introduction of business apps may fall short of improving efficiency. In addition, operating employees have valuable insights into the actual process execution as well as unobvious process exceptions. Giving them a means to shape the software they use in their everyday work routines offers not only the possibility to explicate their tacit knowledge for the development of best practices, but also actively involves them in the evolution of the enterprise. Instead of participating only in early requirements engineering phases of software development, continuously co-designing such systems may increase the adoption of the resulting application and possibly strengthen their identification with the company [19]. Mobile app development can thus benefit from the incorporation of people from all levels of the organization and development tools should be understandable to both programmers and domain experts.

The research company Gartner predicted that more than half of all company-internal mobile apps will be built using codeless tools by 2018 [51]. The general trend towards low-code or codeless development of business apps can be supported by the introduction of graphical notations which are particularly suitable to represent the concepts of a data-driven and process-focused domain. However, current approaches often lack the capacity for holistic app modelling, often operating on a low level of abstraction with visual user interface builders or approaches using view templates (e.g., [22,64]).

In order to advance research in the domain of cross-platform development of mobile apps and investigate opportunities for organizations in a digitized world, this paper presents and evaluates the Münster App Modeling Language (MAML) framework. Rooted in the Eclipse ecosystem, the DSL grammar is defined as an Ecore metamodel, the visual editor is built using the Sirius framework [57], and technologies such as Xtend are used for the code generation of Android, iOS, and Wear OS apps.

Moreover, the high level of abstraction and automatic inference required for simple-to-use app development opens up opportunities for reusing the same notation for heterogeneous target devices. The terms cross-platform or multi-platform development typically denote the creation of applications for multiple platforms *within the same device class*, for example iOS and Android in the smartphone domain – potentially extended to technically similar tablets. However, the development of apps *across device classes* is not yet tackled systematically in academia or practice. To distinguish the additional requirements and challenges introduced when creating applications across heterogeneous devices, we propose the term *pluri-platform development*.

This article greatly extends the paper [46] presented at HICSS 2018[1]. It has been updated to reflect the latest developments, includes new content based on

---

[1] Please note that verbatim content from the original paper is not explicitly highlighted but for figures and tables already included there.

additional work as well as on the discussions at the conference. Also, it has been amended with a perspective on challenges and opportunities regarding model-driven app development for novel device classes and a study on the suitability of the MAML notation in this context. The remainder of this article is structured as follows: After presenting related work in Sect. 2, MAML's graphical DSL is presented that allows for the visual definition of business apps (Sect. 3). The codeless app creation capabilities are demonstrated using the MAML editor with advanced modelling support and an automated generation of native app source code through a two-step model transformation process. Section 4 discusses the setup and results of two usability studies conducted to demonstrate the potential and intricacies of an integrated app modelling approach for a wide audience of process modellers, domain experts, and programmers. The possibility to extend the approach to heterogeneous devices is covered in Sect. 5. In Sect. 6, the findings and implications of MAML are discussed with regard to model-driven development for heterogeneous app-enabled devices before concluding with a summary and outlook in Sect. 7.

2   Related Work

Different approaches to cross-platform mobile app development have been researched. In general, five approaches can be distinguished [35]. Concerning runtime approaches, *mobile webapps* are browser-run web pages optimized for mobile devices but without native user interface (UI) elements, *hybrid approaches* such as Apache Cordova [3] provide a wrapper to web-based apps that allow for accessing device-specific features through interfaces, and *self-contained runtimes* provide separate engines that mimic core system interfaces in which the app runs. In addition, two generative approaches produce native apps, either by *transpiling* apps between programming languages such as J2ObjC [23] to transform Android-based business logic to Apple's language Objective-C, or *model-driven software development* for transforming a common model to code.

   With regard to model-driven development, DSLs are used to model mobile apps on a platform-independent level. According to Langlois et al. [32], DSLs can be classified in textual, graphical, tabular, wizard-based, or domain-specific representations as well as combinations of those. Several frameworks for mobile app development have been developed in the past years, both for scientific and commercial purposes. In the particular domain of business apps – i.e., form-based, data-driven apps interacting with back-end systems [35] – the graphical approach JUSE4Android [15] uses annotated UML diagrams to generate the appearance of and navigation within object graphs, and Vaupel et al. [60] presented an approach focusing around role-driven variants of apps using a visual model representation. Other approaches such as AXIOM [30] and Mobl [26] provide textual DSLs to define business logic, user interaction, and user interface in a common model. An extensive overview of current model-driven frameworks is provided by Umuhoza and Brambilla [59]. However, current approaches mostly rely on a textual specification which limits the active participation of non-technical users without prior training [65], and graphical approaches are often incapable of covering all structural and behavioural aspects of a mobile app. For generating source code, the work in this paper is based on the Model-Driven Mobile Development (MD^2) framework which also uses a textual DSL for specifying all constituents of a mobile app in a platform-independent manner. After preprocessing the models, native source code is generated for each target platform as described by Majchrzak and Ernsting [34]. This intermediate step is, however, automated and requires no intervention by the user (see Subsect. 3.4).

In contrast to DSLs, several general-purpose modelling notations exist for graphically depicting applications and processes, such as the Unified Modeling Language (UML) with a collection of interrelated standards for software development. The Interaction Flow Modeling Language (IFML) can be used to model user interactions in mobile apps, especially in combination with the mobile-specific elements introduced as extension by Breu et al. [11]. Process workflows can for example be modelled using BPMN [40], Event-Driven Process Chains [1], or flowcharts [27]. However, such notations are often either suitable for generic modelling tasks and remain on a superficial level of detail, or represent rather complex technical notations designed for a target group of programmers [18]. A trade-off is necessary to balance the ease of use for modellers with the richness of technical details for creating functioning apps. Moody [38] has pointed out principles for the cognitive effectiveness of visual notations and subsequent studies have revealed comprehensibility issues through effects such as symbol overload, e.g., for the WebML notation preceding IFML [25]. Examples of technical notations in the domain of mobile applications include a UML extension for distributed systems [54] and a BPMN extension to orchestrate web services [10]. Nevertheless, the approach presented in this work goes beyond pure process modelling. While IFML is closest to the work in this paper regarding the purpose of modelling user interactions, MAML covers both structural (data model and views) and behavioural (business logic and user interaction) aspects.

Lastly, visual programming languages have been created for several domains such as data integration [42] but few approaches focus specifically on mobile apps. RAPPT combines a graphical notation for specifying processes with a textual DSL [8], and AppInventor provides a language of graphical building blocks for programming apps [63]. However, non-technical users are usually ignored in the actual development process. Hence, those visual notations do not exploit the potential of including people with in-depth domain knowledge. Considering commercial frameworks, support for visual development of mobile apps varies significantly. In practice, many recent tools are limited to specific components such as back-end systems or content management, or support particular development phases such as prototyping [43]. Start-ups such as Bizness Apps [9] and Bubble Group [13] aim for more holistic development approaches using configurators and web-based editors. Similarly, development environments have started to provide graphical tools for UI development, enhancing the programmatic specification of views by complementary drag and drop editors [64]. The WebRatio Mobile Platform also supports codeless generation of mobile apps through a combination of

IFML, other UML standards, and custom notations [62]. In contrast, this work focuses on a significantly more abstract and process-centric modelling level as presented in the next section.

3   Münster App Modeling Language

At its core, the MAML framework consists of a graphical modelling notation that is described in the following subsections. Contrary to existing notations, its models contain sufficient information to transform them into fully functional mobile apps. The framework also comprises the necessary development tools to design MAML models in a graphical editor and generate apps without requiring manual programming. The generation process is described in more detail in Subsect. 3.4.

3.1   Language Design Principles

The graphical DSL for MAML is based on five design goals:

Automatic cross-platform app creation: Most important, the whole approach is built around the key concept of codeless app creation. To achieve this, individual models need to be recombined and split according to different roles (see Subsect. 3.4) and transformed into platform-specific source code. As a consequence, models need to encode technical information such as data fields and interrelations between workflow elements in a machine-interpretable way as opposed to an unstructured composition of shapes filled with text.

Domain expert focus: MAML is explicitly designed with a non-technical user in mind. Process modellers as well as domain experts are encouraged to read, modify, and create new models by themselves. The language should, therefore, not resemble technical specification languages drawn from the software engineering domain but instead provide generally understandable visualizations and meaningful abstractions for app-related concepts.

Data-driven process modelling: The basic idea of business apps to focus on data-driven processes determines the level of abstraction chosen for MAML. In contrast to merely providing editors for visual screen composition as replacement for manually programming user interfaces, MAML models represent a substantially higher level of abstraction. Users of the language concentrate on visualizing the sequence of data processing steps and the concrete representation of affected data items is automatically generated using adequate input/output user interface elements.

Modularization: To engage in modelling activities without advanced knowledge of software architectures, appropriate modularization is important to handle the complexity of apps. MAML embraces the aforementioned process-oriented approach by modelling use cases, i.e., a unit of functionality containing a self-contained set of behaviours and interactions performed by the app user [41]. Combining data model, business logic, and visualization in a single model deviates from traditional software engineering practices which, for instance, often rely on the Model-View-Controller pattern [20]. In accordance with the domain expert focus, the end user is, however, unburdened from this technical implementation issue.

Declarative description: MAML models consist of platform-agnostic elements, declaratively describing *what* activities need to be performed with the data. The concrete representation in the resulting app is deliberately unspecified to account for different capabilities and usage patterns of each targeted mobile platform. The respective code generator can provide sensible defaults for such platform specifics.

3.2   Language Overview

In the following, the key concepts of the MAML DSL are highlighted using the fictitious scenario of a publication management app. A sample process to add a new publication to the system consists of three logical steps: First, the researcher enters some data on the new publication. Then, he can upload the full-text document and optionally revise the corresponding author information. This self-contained set of activities is represented as one model in MAML, the so-called use case, as depicted in Fig. 1.

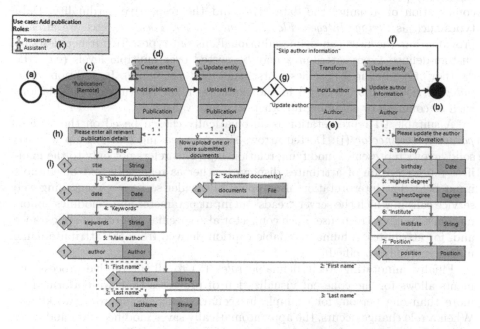

Fig. 1. MAML use case for adding a publication to a review management system [45]

A model consists of a *start event* (labelled with (a) in Fig. 1) and a sequence of process flow elements towards an *end event* (b). A *data source* (c) specifies what type of entity is first used in the process, and whether it is only saved locally on the mobile device or managed by the remote back-end system. Then, the modeller can choose from predefined *interaction process elements* (d), for example to *create/show/update/delete* an entity, but also to *display messages*, access device sensors such as the *camera*, or *call* a telephone number. Because of the declarative description, no device-specific assumptions can be made on the appearance of such a step. The generator instead provides default representations and functionalities, e.g., display a *select entity* step using a list of all available objects as well as possibilities for searching or filtering. In addition, *automated process elements* (e) represent steps to be performed without user interaction. Those elements provide the minimum amount of technical specificity in order to navigate between the model objects (*transform*), request information from *web services*, or *include* other models to reuse existing use cases.

The order of process steps is established using *process connectors* (f), represented by a default "Continue" button unless specified differently along the connector element. *XOR* (g) elements branch out the process flow based on a manual user action by rendering multiple buttons (see differently labelled connectors in Fig. 1), or automatically by evaluating expressions referring to a property of the considered object.

The lower section of Fig. 1 contains the data linked to each process step. *Labels* (h) provide explanatory text on screen. *Attributes* (i) are modelled as combination of a name, the data type, and the respective cardinality. Data types such as *String, Integer, Float, PhoneNumber, Location*, etc. are already provided but the user can define additional custom types. To further describe custom-defined types, attributes may be nested over multiple levels (e.g., the "author" type in Fig. 1 specifies a first name and last name). In addition, *computed attributes* (not depicted in the example) allow for runtime calculations such as counting or summing up other attribute values.

A suitable UI representation is automatically chosen based on the type of *parameter connector* (j): Dotted arrows signify a reading relationship whereas solid arrows represent a modifying relationship. This refers not only to the manifest representation of attributes displayed either as read-only text or editable input field. The interpretation also applies in a wider sense, e.g., regarding web service calls in which the server "reads" an input parameter and "modifies" information through its response. Each connector also specifies an order of appearance and, for attributes, a human-readable caption derived from the attribute name unless manually specified.

Finally, annotating freely definable *roles* (k) to all interactive process elements allows for the coherent visualization of processes that are performed by more than one person, for example in scenarios such as approval workflows. When a role change occurs, the app automatically saves modified data and users with the subsequent role are informed about the open workflow instance in their app.

3.3   App Modelling

In contrast to other notations, all of the modelling work is performed in a single type of model, mainly by dragging elements from a palette and arranging them on a large canvas. The modelling environment was developed using the Eclipse Sirius framework [57] that was extended with domain-specific validation and guidance to provide advanced modelling support for MAML.

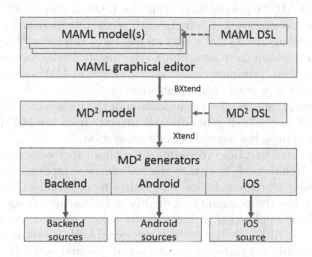

Fig. 2. MAML app generation process (cf. [46])

Modelling only the information displayed in each process step effectively creates a multitude of partial data models for each process step and for each use case as a whole. Also, attributes may be connected to multiple process elements simultaneously, or can be duplicated to different positions to avoid wide-spread connections across the model. An inference mechanism [45] aggregates and validates the complete data model while modelling. During generation, app-internal and back-end data stores are automatically created. As a result, the user does not need to specify a distinct global data model and consistency is automatically checked when models change.

Apart from validation rules to prevent users from modelling syntactically incorrect MAML use cases in the first place, additional validity checks have been implemented in order to detect inconsistencies across use cases (based on the inferred data model) as well as potentially unwanted behaviour (e.g., missing role annotations). Moreover, advanced modelling support attempts to provide guidance and overview to the user. For example, the current data type of a process element (lower label of (d) in Fig. 1) is automatically derived from the preceding elements to improve the user's imagination within the process. Also, suggestions of probable values are provided when adding elements (e.g., known attributes of the originating type when adding UI elements).

3.4   App Generation

Technically, MAML relies on and integrates with the Eclipse Modeling Framework (EMF), for example by specifying the DSL's metamodel as an Ecore model. In order to generate apps, the proposed approach reuses previous work on MD^2 (see Sect. 2). The complete generation process is depicted in Fig. 2. Because of space constraints, the respective transformations are only sketched next.

First, model transformations are applied to transform graphical MAML models to the textual MD^2 representation using the BXtend framework [14] and the Xtend language. Amongst other activities, all separately modelled use cases are recombined, a global data model across all use cases is inferred and explicated, and processes are broken down according to the specified roles. In the subsequent code generation step, previously existing generators in MD^2 create the actual source code for all supported target platforms.

This is, however, not an inherent limitation of the framework. Newly created generators might just as well generate code directly from the MAML model or use interpreted approaches instead of code generation.

It should be noted that this proceeding differs from approaches such as UML's Model Driven Architecture [5] in that the intermediate representation is still a platform-independent representation but with a more technical focus. Optionally, a modeller has the possibility to modify default representations and configure parts of the application in more detail before source code is generated for each platform. Although the tooling around MAML is still in a prototypical state, it currently supports the generation of Android and iOS apps as well as a Java-based server back-end component. Also, a smartwatch generator for Google's Wear OS platform highlights the applicability to further device classes (cf. Sect. 5). The screenshots in Fig. 3 depict the generated Android app views for the first process steps of the MAML model depicted in Fig. 1.

4   Evaluation

As demonstrated, MAML aligns with the goals of automated cross-platform app creation from modular and platform-agnostic app models (cf. Subsect. 3.1). However, the suitability of data-driven process models with regard to the target audience needed to be evaluated in more detail. Therefore, an observational study was performed to assess the utility of the newly developed language. After describing the general setup in Subsect. 4.1, the results on comprehensibility and usability of the graphical DSL are presented.

4.1   Study Setup

The purpose of the study was to assess MAML's claim to be understandable and applicable by users with different backgrounds, in particular including non-technical users. From the variety of methodologies for usability evaluation, observational interviews according to the think-aloud method were selected as empirical approach [21]. Participants were requested to perform realistic tasks with the system under test and urged to verbalize their actions, questions, problems, and general thoughts while executing these tasks. Due to the novelty of MAML which excludes the possibility of comparative tests, this setup focused on obtaining detailed qualitative feedback on usability issues from a group of potential users.

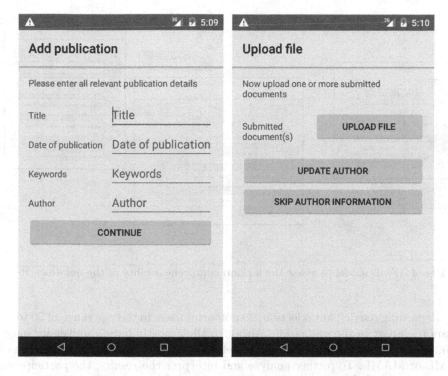

Fig. 3. Exemplary screenshots of generated Android app views [46]

Therefore, 26 individual interviews of around 90 min duration were conducted. An interview consisted of three main parts: First, an online questionnaire had to be filled out in order to collect demographic data, previous knowledge in the domains of programming or modelling, and personal usage of mobile devices. Second, a MAML model and an equivalent IFML model were presented to the participants (in random order to avoid bias) to assess the comprehensibility of such models without prior knowledge or introduction. In addition to the verbal explanations, a short 10-question usability questionnaire was filled out to calculate a score according to the System Usability Scale (SUS) [12] for each notation (cf. Subsect. 4.2). Third, the main part of the interview consisted of four modelling tasks to accomplish using the MAML editor. Finally, the standardized ISONORM questionnaire was used to collect more quantitative feedback, aligned with the seven key requirements of usability according to DIN 9241/110-S [28] (cf. Subsect. 4.3).

To capture the variety of possible usability issues, 71 observational features were identified a priori and structured in six categories of interest: comprehensibility, applying the notation, integration of elements, tool support, effectiveness, and efficiency. In total, over 1500 positive or negative observations were recorded as well as additional usability feedback and proposals for improvement.

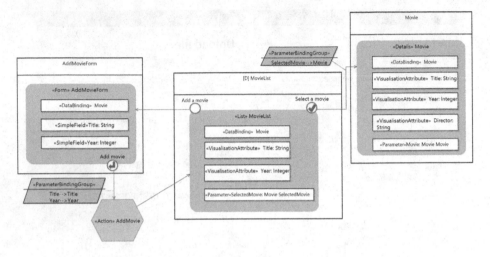

Fig. 4. IFML model to assess the a priori comprehensibility of the notation [46]

Regarding participant selection, 26 potential users in the age range of 20 to 57 years took part in the evaluation. Although they mostly have a university background, technical experience varied widely and none had previous knowledge of IFML or MAML. To further analyse and interpret the results, the participants were categorized in three distinct groups according to their personal background stated in the online questionnaire: 11 software developers have at least medium knowledge in traditional/web/app programming or data modelling, 9 process modellers have at least medium knowledge in process modelling (exceeding their programming skills), and 6 domain experts are experienced in the modelling domain but have no significant technical or process knowledge. Although it is debated whether Virzi's [61] statement of five participants being sufficient to uncover 80% of usability problems in a particular software holds true [56], arguably the selected amount of participants in this study is reasonable with regard to finding the majority of grave usability defects for MAML and generally evaluating the design goals.

For their private use, participants stated an average smartphone usage of 19.2 h per week, out of which 16.3 h are spent on apps. In contrast, tablet use is rather low with 3.5 h (3.2 h for apps), and notebook usage is generally high with 27.5 h but only 4.7 h are spent on apps. For business uses, similar patterns can be observed on total/app-only usage per week on smartphones (5.5h/4.3h), tablets (0.7h/0.2h), and notebooks (18.2h/3.7h). Although this sample is too low for generalizable insights, the figures indicate a generally high share of app usage on smartphones and tablets compared to the total usage duration, both for personal and business tasks. In addition, with mean values of 1.81/2.12 on a scale between 0 (strongly reduce) and 4 (strongly increase), the participants stated to have no desire of significantly changing their usage volumes of private/business apps.

4.2 Comprehensibility Results

Before actively introducing MAML as modelling tool, the participants should explicate their understanding of a given model without prior knowledge. Comprehensibility is an important characteristic in order to easily communicate app-related concepts via models without the need for extensive training. To compare the results with an existing modelling notation, equivalent IFML (see Fig. 4) and MAML models [47] of a fictitious movie database app were provided with the task to describe the purpose of the overall model and the particular elements. The monochrome models were shown to the participants on paper in randomized order to avoid bias from priming effects [7] and potential influences from a particular software environment.

After each model, participants were asked to answer the SUS questionnaire for the particular notation. This questionnaire has been applied in many contexts since its development in 1986 and can be seen as easy, yet effective, test to determine usability characteristics. Each participant answers ten questions using a five-point Likert-type scale between strong disagreement and strong agreement, which is later converted and scaled to a [0;100] interval according to Brooke [12]. The participants' scores for both languages and the respective standard deviations are depicted in Table 1.

Table 1. System usability scores for IFML and MAML

SUS ratings	IFML	MAML
All participants	52.79 ($\sigma = 23.0$)	66.83 ($\sigma = 15.6$)
Software developers	45.91 ($\sigma = 23.6$)	64.09 ($\sigma = 17.3$)
Process modellers	64.17 ($\sigma = 19.0$)	69.44 ($\sigma = 12.0$)
Domain experts	48.33 ($\sigma = 24.5$)	67.92 ($\sigma = 18.7$)

However, it should be noted that the results do not represent percentage values. Instead, an adjective rating scale was proposed by Bangor et al. [6] to interpret the results as depicted in Fig. 5. The results show that MAML's scores are superior overall as well as for all three groups of participants. In addition, the consistency of scores across all groups supports the design goal of creating a notation which is well understandable for users with different backgrounds. Particularly, domain experts without technical experience expressed a drastic difference in comprehensibility of almost 20 points.

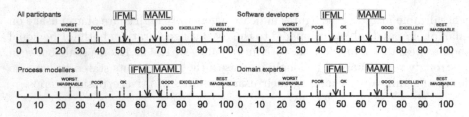

Fig. 5. SUS ratings for IFML and MAML [46]

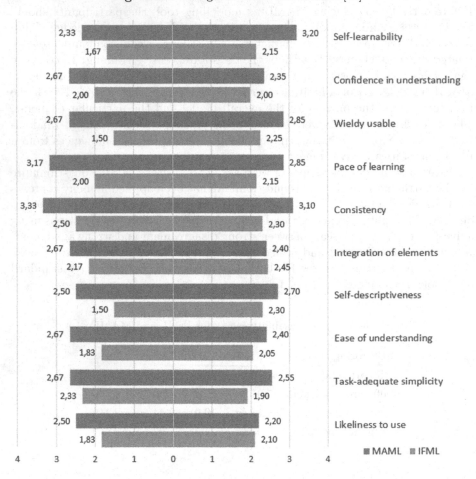

Fig. 6. SUS answers for domain experts (left) and technical users (right) [48]

Especially the distinction between domain experts and technical users (developers and process modellers together) is of interest to evaluate the design goal of MAML to be comprehensible for different user groups. Figure 6 breaks down the answers to the 10 questions of the SUS questionnaire (rescaled to a [0;4] interval; 4 denoting strong acceptance). With one exception, responses for MAML are higher than for the technical IFML notation. Moreover, domain experts reacted significantly more positively when assessing the MAML notation as being wieldy usable (+1.17 compared to IFML), fast to learn (+1.17), and self-descriptive (+1.00). Consequently, the understandability and general applicability of the notation is in the focus of domain experts, which aligns well with the intention to use MAML for communicating with potential end users and include them in the development process. The strongest deviations for technical users, in contrast, can be seen in questions regarding self-learnability (+1.05), perceived consistency (+0.80), and pace of learning (+0.70). Conforming with their technical background, these aspects emphasize the correct application of the notation which apparently is perceived as positive in MAML, too.

Considering also the qualitative observations, some interesting insights can be gained. According to the questionnaire results, most of the criticism is related to the categories "easy to understand" and "confidence in the notation". IFML's approach of visually hinting at the outcome through the order of elements and their composition in screen-like boxes was often noted as positive and slightly more intuitive compared to MAML. This argument is not unexpected as the level of abstraction was designed to be higher than a pure visual equivalent of programming activities. Also, the notation is not limited to the few types of mobile devices known by a participant, e.g., smartwatches and smartphones exhibit very different interface and interaction characteristics. Therefore, a fully screen-oriented approach generally contradicts the desired platform-independent design of MAML. However, this is valuable feedback for the future, e.g., improving modelling support by using an additional simulator component to preview the outcome while modelling.

Surprisingly, IFML scores were worst for the group of software developers, although they have knowledge of other UML concepts and diagrams. Despite this apparent familiarity, reasons for the negative assessment of IFML can be found in the amount of "technical clutter", e.g., regarding parameter and data bindings, as well as perceived redundancies and inconsistencies. In contrast, 86% of these participants highlight the clarity of MAML regarding the composition of individual models and 88% are able to sketch a possible appearance of the final app result based on the abstract process specification.

Overall, three in four participants can also transfer knowledge from other modelling notations, e.g., to interpret elements such as data sources. All participants within the process modeller group immediately recognize analogies from other graphical notations such as BPMN, and understand the process-related concepts of MAML. Whereas elements such as data sources (understood by 75% of all participants) and nested attribute structures (83%) are interpreted correctly on an abstract level, comprehensibility drops with regard to technical aspects, e.g., data types (57%) or connector types (43%).

Finally, domain experts also have difficulties to understand the technical aspects of MAML without previous introduction. Although concepts such as cardinalities (0%), data types (25%), and nested object structures (67%) are not initially understood and ignored, all participants in this group are still able to visualize the process steps and main actions of the model. As described in Subsect. 3.1, further reducing these technical aspects constrains the possibilities to generate code from the model. Some suggestions exist to improve readability, e.g., replacing the textual data type names with visualizations. Nevertheless, MAML is comparatively well understandable. Curiously enough, the sample IFML model is often perceived as being a more detailed technical representation of MAML instead of a notation with equivalent expressiveness.

To sum up, MAML models are favoured by participants from all groups, despite differences in personal background and technical experience. This part of the study is not supposed to discredit IFML but emphasizes their different foci: Whereas IFML covers an extensive set of features and integrates into the UML ecosystem, it is originally designed as generic notation for modelling user interactions and targeted at technical users. In contrast, the study confirms MAML's design principle of an understandable DSL for the purpose of mobile app modelling.

4.3   Usability Results

In addition to the language's comprehensibility, a major part of the study evaluated the actual creation of models by the participants using the developed graphical editor. After a brief ten-minute introduction of the language concepts and the editor environment, four tasks were presented that cover many of MAML's features and concepts. In the hands-on context of a library app (cf. supplementary online material [47]), a first simple model to add a new book to the library requires the combination of core features such as process elements and attributes. Second, participants should model how to borrow a book based on screenshots of the resulting app. This requires more interaction element types, a case distinction, and complex attributes. Third, modelling a summary of charges includes a web service call, exception handling, and calculations. Fourth, a partial model in a multi-role context needed to be altered.

The final evaluation was performed using the ISONORM questionnaire in order to assess the usability according to the ISO 9241-110 standard [28]. 35 questions with a scale between −3 and 3 cover the seven criteria of usability as presented in Table 2. Again, MAML achieves positive results for every criterion, both for the participant subgroups and in total. Taking the interview observations into account for qualitative feedback, these figures can be evaluated in more detail.

Regarding the *suitability for the task*, observations on the effectiveness and efficiency of the notation show that handling models in the editor is achieved without major problems. 94% of the participants themselves noticed a fast familiarization with the notation, although domain experts are generally more wary when using the software. The deliberately chosen high level of abstraction manifests in 37% of participants describing this approach as uncommon or astonishing (see also Sect. 6). Nevertheless, 67% of the participants state to have an understanding of the resulting app while modelling.

Table 2. ISONORM usability questionnaire results for MAML.

Criterion	All participants	Software developers	Process modellers	Domain experts
Suitability for the task	1.63 ($\sigma = 1.04$)	1.36 ($\sigma = 1.13$)	1.62 ($\sigma = 1.12$)	2.13 ($\sigma = 0.62$)
Self-descriptiveness	0.51 ($\sigma = 0.73$)	0.62 ($\sigma = 0.62$)	0.38 ($\sigma = 1.02$)	0.50 ($\sigma = 0.41$)
Controllability	2.10 ($\sigma = 0.83$)	2.20 ($\sigma = 0.63$)	2.02 ($\sigma = 0.63$)	2.03 ($\sigma = 1.41$)
Conformity with user expectations	1.78 ($\sigma = 0.52$)	1.85 ($\sigma = 0.47$)	1.64 ($\sigma = 0.47$)	1.87 ($\sigma = 0.70$)
Error tolerance	0.92 ($\sigma = 0.96$)	0.89 ($\sigma = 0.63$)	1.11 ($\sigma = 0.81$)	0.70 ($\sigma = 1.63$)
Suitability for individualisation	1.20 ($\sigma = 0.90$)	1.04 ($\sigma = 1.05$)	1.42 ($\sigma = 1.02$)	1.17 ($\sigma = 0.27$)
Suitability for learning	1.83 ($\sigma = 0.67$)	2.02 ($\sigma = 0.54$)	1.69 ($\sigma = 0.66$)	1.70 ($\sigma = 0.90$)
Overall score	1.43 ($\sigma = 0.49$)	1.43 ($\sigma = 0.46$)	1.41 ($\sigma = 0.53$)	1.44 ($\sigma = 0.59$)

Self-descriptiveness refers to comprehension issues but additionally deals with the correct integration of different elements while modelling. For example, the concept of user roles was introduced to the participants but not the assignment in models. Still, 86% of them intuitively drag and drop role icons on process elements correctly. Furthermore, process exceptions were not explained at all in the introduction but 71% of the participants applied the "error event" element correctly without help. Self-descriptiveness is, however, more limited when dealing with technical issues. Side effects of transitive attributes are only recognized by 43% of process modellers and 25% of domain experts. Model validation or additional modelling support is needed in order to guide the users towards semantically correct models. Similarly, the complexity of modelling web service responses within the use case's data flow poses challenges to 44% of the participants.

The very positive responses for the *controllability* criterion can be explained by the simplistic design of MAML and its tools. All modelling activities are performed in a single model instead of switching between multiple perspectives. In contrast to other notations, all of the modelling work is performed in a single type of model, mainly by dragging elements from a palette and arranging them on a large canvas. Many participants utter remarks such as "the editor does not evoke the impression of a complex tool". In parts, this impression can be attributed to sophisticated modelling support, including live data model inference when connecting elements in the model, validation rules, and suggestions for available data types.

Related to the clarity of possible user actions, the *conformity with user expectations* is also clearly positive. Despite occasional performance issues caused by the prototypical nature of the tools, a consistent handling of the program is confirmed by the participants. Although aspects such as the direction of parameter connections may be interpreted differently (e.g., either a sum refers to attributes or attributes are incoming arguments to the sum function), the consistent use of concepts throughout the notation is easily internalized by the participants.

Regarding *error tolerance* and *suitability for individualisation*, scores are moderate but the prototype was not yet particularly optimized for production-ready stability or performance. Also, an individual appearance was not intended, thus providing only basic capabilities such as resizing and repositioning components. Whereas the editor is very permissive with regard to the order of modelling activities, adding invalid model elements is mostly avoided by syntactic and semantic validity checks, e.g., which elements are valid end points of a connector. Participants appreciate the support of not being able to model invalid constellations. However, criticism arises from disallowing actions without further feedback on why a specific action is invalid. The modelling environment Sirius is currently not able to provide this information, yet users might benefit more from such dynamic explanations than from traditional help pages.

Finally, *suitability for learning* can be demonstrated best using quotes such as MAML being judged as "a really practical approach", and participants having "fun to experiment with different elements" or being "surprised about what I am actually able to achieve". Using the graphical approach, users can express their ideas and apply concepts consistently to different elements. As mentioned above, many unknown features such as roles or web service interactions can be learned using known drag and drop patterns or read/modify relationships.

5 Towards Pluri-Platform Development

The term *cross-platform* as well as actual development frameworks in academia and practice are usually limited to smartphones and sometimes – yet not always – technically similar tablets. Thus, they ignore the differing requirements and capabilities within the variety of novel devices and platforms reaching the mainstream consumer market in the near future. Extending the boundaries of current cross-platform development approaches requires a new scope of target devices which can be subsumed under the term *app-enabled* devices. Following the definition in [50], an app-enabled device can be described as being extensible with software that comes in small, interchangeable pieces, which are usually provided by third parties unrelated to the hardware vendor or platform manufacturer, and increase the versatility of the device after its introduction. Although these devices are typically portable or wearable and therefore related to the term *mobile computing*, there are further devices classes with the ability to run apps (e.g., smart TVs).

However, new challenges arise when extending the idea of cross-platform development to app-enabled device classes. In particular, not all approaches mentioned in Sect. 2 are generally capable for this extension as described in the following.

5.1 Challenges

From the development and usage perspectives on app development, specific challenges can be identified related to app development across device classes and which can be grouped into four main categories.

Output Heterogeneity: The user interface of upcoming device classes enables more flexible and intuitive ways of device interaction compared to the prevalent focus on medium screen sizes between 4″ and 10″. By design, graphical output is very limited on wearables. On the other hand, smart TVs provide large-scale screens beyond 20″. Also, devices can use new techniques for presenting information, e.g., auditive output by smart virtual assistants, or projection through augmented reality (for example using the wind shield in vehicles). In addition, even for screen-based devices the variability of output increases because of new screen designs with drastically differing pixel density, aspect ratios, and form factors (e.g., round smartwatches) [50]. Techniques from the field of adaptive user interfaces may be used to tackle these issues. To achieve this degree of adaptability, specifying user interfaces needs to evolve from a screen-oriented specification of explicitly positioned widgets to a higher level of abstraction which can use semantic information to transform the content to a particular representation.

User Input Heterogeneity: The device interfaces for entering information by the user also evolve and will use a wider spectrum of possible techniques for user input [50]. This ranges from pushing buttons attached to the device, using remote controls, directing pointing devices for graphical user interfaces, tapping on touch screens, and using auxiliary devices (e.g., stylus pens) to hands-free interactions via gestures, voice, or even neural interfaces. To complicate matters, a single device may provide multiple input alternatives for convenience and especially new device classes are often experimenting with different interactions patterns. Again, this complexity calls for a higher level of abstraction when specifying apps by decoupling actual input events from the intended actions of the user interaction.

Device Class Capabilities: The variability of hardware and software across device classes is also apparent besides the user interface. For example, the miniaturization in wearable devices negatively impacts the computational power and battery capacity. Complex computations may therefore be offloaded to potential companion devices or provided through edge/cloud computing [44]. Sensing capabilities can vary both within and across device classes. In addition, platform operating systems provide different levels of device functionality access and app interoperability, e.g., regarding security issues in vehicles. To avoid the problem of developing for the least common denominator of all targeted devices, suitable replacements for unavailable sensors need to be provided. For example, automatic location detection via GPS sensors can have fallback solutions such as address lookup or manual selection on a map.

Multi-device Interaction: Whereas cross-platform approaches often provide self-contained apps with the same functionality for different users (it is fairly uncommon to own multiple smartphones with different platforms), users increasingly own multiple devices of different device classes and aim for interoperable solutions within their ecosystem. This complexity of multi-device interactions for a single user might occur *sequentially* when a user switches to a different device depending on the usage context or user preferences (e.g., reading notifications on a smartwatch and typing the response on a smartphone for convenience). Moreover, a *concurrent* usage of multiple devices for the same task is possible, for instance in a second screening scenario in which one device provides additional information or input/output capabilities for controlling another device [39]. In both cases, fast and reliable synchronization of content is essential in order to seamlessly switch between multiple devices.

5.2   Towards Pluri-Platform Development

To emphasize the difference in scope and the respective solution approaches compared to traditional cross-platform development, we propose the term *pluri-platform* development to signify the creation of apps *across* device classes, in contrast to multi-/cross-platform development for several platforms *within* one class of mostly homogeneous devices. Pluri-platform development can, therefore, be understood as an umbrella term for different approaches aiming to bridge the gap between multiple device classes by tackling the challenges of heterogeneous input and output mechanisms, device capabilities, and multi-device interactions. In contrast to cross-platform development, the focus lies on simplification of app creation not just with regard to the representation of user interfaces but also the integration with platform-specific usage patterns and the interaction within a multi-device context. The related research fields of adaptive user interfaces and context-aware interfaces thus only account for a subset of the required solutions to achieve pluri-platform development.

Considering previous literature in this domain, very few works explicitly deal with app development *spanning multiple device classes*, indicating that app development beyond smartphones is not yet approached systematically but on a case-by-case basis. Cross-platform overview papers such as [29] typically focus on a single category of devices and apply a very narrow notion of mobile devices. [50] provides the only classification that includes novel device classes. Few papers provide a *technical* perspective on apps spanning multiple device classes. Singh and Buford [55] describe a cross-device team communication use case for desktop, smartphones, and wearables, and Esakia et al. [17] performed research on the interaction between the Pebble smartwatch and smartphones in computer science courses. In the context of Web-of-Things devices, Koren and Klamma [31] propose a middleware approach to integrate data and heterogeneous UI, and Alulema et al. [2] propose a DSL for bridging the presentation layer of heterogeneous devices in combination with web services for incorporating business logic.

With regard to commercial cross-platform products, Xamarin [64] and CocoonJS [33] provide Wear OS support to some extent. Whereas several other frameworks claim to support wearables, this usually only refers to accessing data by the main smartphone application or displaying notifications on already coupled devices.

Together with the increase in devices, new software platforms have appeared, some of which are either related to established operating systems (OS) for other device classes or are newly designed to run on multiple heterogeneous devices. Examples include Android/Wear OS, watchOS, and Tizen. Although these platforms ease the development of apps (e.g., reusing code and libraries), subtle differences exist in the available functionality and general cross-platform challenges remain.

5.3 Applicability of Existing Cross-Platform Approaches for Pluri-Platform Development

Different instantiations and practical frameworks may be conceived which extend the approaches to cross-platform development presented in Sect. 2.

A plethora of literature exists in the context of cross- or multi-platform development. Classifications such as in [16] and [36] have identified five main approaches to multi-platform app development which are varyingly suited to the specific challenges of pluri-platform development. With regard to runtime-based approaches, *mobile webapps* – including recently proposed Progressive Web Apps – are mobile-optimized web pages that are accessed using the device's browser and relatively easy to develop using web technologies. However, most novel device types, e.g., the major smartwatch platforms watchOS [4] and Wear OS by Google [24], do not provide WebView components or browser engines that allow for the execution of JavaScript code. Consequently, this approach cannot be used for pluri-platform development targeting a broader range of devices.

Hybrid apps are developed similarly using web technologies but are encapsulated in a wrapper component that enables access to device hardware and OS functionality through an API. Although they are distributed as app packages via marketplaces, hybrid apps rely on the same technology and can neither be used for pluri-platform development.

In contrast, a *self-contained runtime* does not depend on the device's browser engine but uses platform-specific libraries provided by the framework developer in order to use native functionality. Of the runtime environment approaches, this is the only one that can be used for pluri-platform development. Although usually based on custom scripting languages, a runtime can also be used as a replacement for inexistent platform functionality. As an example, CocoonJS [33] recreated a restricted WebView engine and, therefore, supports the development of JavaScript-based apps also for the Wear OS platform. However, devices need sufficient computing power to execute the runtime on top of the actual operating system. Also, synergies with regard to user input/output and available hardware/software functionality across heterogeneous devices are dependent on the runtime's API.

Considering generative approaches to cross-platform development, *model-driven software development* has several advantages as it uses textual or graphical models as main artefacts to develop apps and then generates native source code from this platform-neutral specification. Referring to domain-specific concepts allows for a high level of abstraction, for example circumventing issues such as input and output heterogeneity using declarative notations. Arbitrary platforms can be supported by developing respective generators which implement a suitable mapping from descriptive models to native platform-specific implementations.

Finally, *transpiling* approaches use existing application code and transform it into different programming languages. Pluri-platform development using this approach is technically possible as the result is also native code. However, there is more to app development than just the technical equivalence of code, which also explains the low adoption of this approach by current cross-platform frameworks. For instance, user interfaces behave drastically differently across different device classes, and substantial transformations would be required to identify the contextual patterns from the low-level implementation. It is therefore unlikely that this approach provides a suitable means for pluri-platform development beyond reusing individual components such as business logic.

To sum up, only self-contained runtimes and model-driven approaches are candidates for pluri-platform development, of which the latter additionally benefits from the transformation of domain abstractions to platform-specific implementations.

5.4   Evaluation of MAML in a Pluri-Platform Context

The MAML notation condenses the development of apps to a sequence of data manipulation activities. Conceptually, the platform-agnostic nature and high level of abstraction allow for a wider applicability beyond just smartphone platforms. From the variety of novel app-enabled device classes [50], smartwatches have so far become most prevalent on the consumer market which offers a multitude of devices and several vendors promoting new platforms. Today's situation resembles the early experimental years after the introduction of the iPhone in 2007 and development using adequate abstractions is needed. Although a smartwatch typically has a touch screen, the screen dimension as well as the user input mechanisms, sensing capabilities, and usage patterns differ from smartphones. Ideally, pluri-platform development approaches can bridge the gap between app development not only for different smartwatches but integrate with existing ecosystems of current app-enabled devices. This is especially beneficial in a multi-device context in which one user owns several devices and can use platform-adapted apps depending on personal preferences or usage contexts.

To investigate the practical opportunities and challenges of pluri-platform development, we developed a new code generator for the Wear OS platform by Google (formerly Android Wear) [24] which supports the creation of stand-alone apps for respective smartwatches. Consequently, the model-driven foundation of our framework now allows for the combined generation of smartphone and smartwatch source code using the same MAML models as input. More details on the required transformations to represent the desired content on a smartwatch are presented in [49]. Yet, in this work we want to focus on the usability of the notation with regard to the specification of apps across device classes.

Therefore, a second study was conducted in order to validate the previous results in the established smartphone domain and gain insights on the suitability towards other app-enabled device classes. The study was conducted with 23

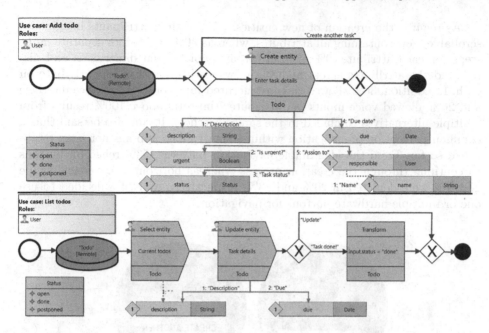

Fig. 7. Use cases for adding and displaying items in a to-do management system [45]

students from a course on advanced concepts in software engineering of an information systems master's program. Whereas designing applications using MDSD techniques is part of the course contents and knowledge of process modelling notations can be presumed, no previous experience with app development was expected in order to avoid a bias towards existing frameworks or approaches. This is supported by the average responses regarding experience in the development of web apps (3.26), hybrid apps (4.35), and native apps (4.30) on a 5-point Likert scale.

Using a simple to-do management scenario depicted in Fig. 7, a 5-min introduction to the MAML notation was given to explain the two processes of creating a new to-do item in the system and displaying the full list of to-dos with the possibility to update items and ticking off the task. Subsequently, participants were asked to express their conceptions of the resulting apps by sketching smartphone and smartwatch user interfaces complying with these use cases.

Interestingly, 64% of the participants intuitively chose a square representation for the smartwatch screen, which reflects the publicity of the Apple watch. Also, a variety of interaction patterns could be derived from the sketches, for example the representation of repetitive elements as a vertical scrollable list (65%) in contrast to 17% using a horizontal arrangement. From the sketches that hint towards navigation patterns, the master-detail pattern of the "list todos" use case of Fig. 7 was conceived either via tapping on the element (42%), using an edit button (33%), pressing a hardware button such as the watch crown (8%), swiping to the right hand side (8%), or using a voice command (8%).

As regards the creation of new entities, 35% of the participants imagined a scrollable view containing all attributes, whereas 30% decided for separate input steps for each attribute, 9% utilized the available screen dimensions and distributed the attributes across multiple views with more than one attribute on each. In addition, 30% allowed the unstructured input of data via voice interface and 26% allowed voice inputs per attribute (the total above 100% results from multiple alternatives combined in the same sketches). It can also be said that a common perception of navigation within a smartphone app has not been established so far: From the identifiable navigation patterns, 53% relied on buttons to continue through the creation process whereas horizontal or vertical swipe gestures were depicted in 20% and 13%, respectively, and 14% decided to use one or multiple hardware buttons for navigation.

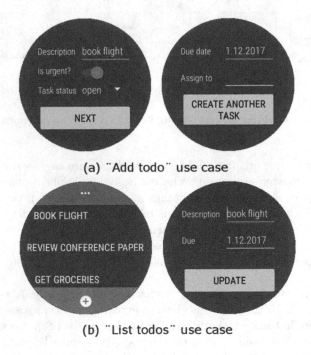

(a) "Add todo" use case

(b) "List todos" use case

Fig. 8. Generated Wear OS app for the system modelled in Fig. 7 (cf. [49])

The standardized SUS questionnaire was used to triangulate the results with the initial study presented in Sect. 4. The resulting score of 66.85 ($\sigma = 12.9$) aligns very well with the figures depicted in Table 1 for software developers and process modellers and reinforces the validity of the previous study. Upon showing the generated app result depicted in Fig. 8, the participants were asked about their opinion on the smartwatch outcome (using again a 5-point Likert scale). The participants agreed (2.04) that the generated smartwatch app suitably represents the process depicted in the MAML model. Furthermore, they supported the statement (2.39) that the resulting app is functional with regard to the to-do management scenario. The visual appearance of the smartwatch was rated merely with 3.3 which can be explained by the generic transformations and assumptions derived from the abstract process model. Also, the prototypical nature of our generator needs more refinements to choose suitable representations.

Regarding the combined generation of apps for smartwatch and smartphone from the same model, the participants did not feel that the common notation makes app development unnecessarily complex (3.35) and tended to agree that having one notation for both app representations accelerates app development (2.48). When asked about specific durations, the students estimated the required time to build the MAML models with 50 min on average, compared to a mean value of 27.3 h when programming the application natively or with cross-platform programming frameworks. Though the actual development was not performed in this study, these estimates underline the possible economic impact of MDSD to reduce the effort for creating specific applications and thus achieve a faster time to market for new apps or app updates.

6 Discussion

In this section, key findings of the proposed MAML framework and subsequent evaluation are discussed with regard to the design objectives and general implications on model-driven software development for mobile applications across device classes. Regarding the principle of data-driven process modelling, using process flows in a graphical notation has shown to be a suitable approach for declaratively designing business apps. Graphical DSLs can also simplify modelling activities for the users of other domains, especially those that benefit from a visual composition of elements such as graph structures. Particularly for MAML, the chosen level of abstraction allows for a much wider usage compared to low-level graphical screen design: Besides the actual app product, models can be used to discuss and communicate small-scale business processes in a more comprehensive way than BPMN or similar process notations through combined modelling of process flows and data structures. In contrast to alternative codeless app development approaches focused on the graphical configuration of UI elements, users do not get distracted by the eventual position of elements on screen but can focus on the task to be accomplished. Moreover, the DSL is platform-agnostic and can thus be used to describe apps for a large variety of mobile devices. Apart from smartphones and tablets, generators for novel device types such as smartwatches or smart glasses may be created in the future based on the same input models.

Second, the challenge of developing a machine-interpretable notation that is understandable both for technical and non-technical users is a balancing act, but the interview observations and consistent scores in the evaluation indicate this design goal was reached. The most significant differences in the participants' modelling results are related to technical accuracy, mostly because of (missing) knowledge about programming or process abstractions. As such issues not always manifest as modelling errors but often happen through oversights, preventing them while keeping a certain joy of use is only achievable using a combined approach: The notation itself should be permissive instead of overly formal. Moreover, clarity (e.g., wording of UI elements) and simplicity of the DSL contribute to manageable models. Most important, however, is the extensive use of modelling support for different levels of experience. Novice users learn from hints (e.g., hover texts and error explanations) whereas advanced users can benefit from domain-specific validation rules and optional perspectives to preview results of model changes. Particularly for MAML, advanced modelling support is achieved by interpreting the models and inferring a global object structure from a variety of partial data models as described in [45]. Consequently, this feature allows for dynamically generated suggestions such as available data types, implicit reactions such as forbidding illegitimate element connections, and validation of conflicting data types and cardinalities. In general, a model-driven approach with advanced modelling support enables the active involvement of business experts in software development processes and can be regarded as major influencing factor for a successful integration of non-programmers.

Finally, the choice of mixing data model, business logic, and view details in a single model deviates from traditional software engineering practices in order to ease the modelling process for non-technical users. This does not mean that we recommend MAML for all process-oriented modelling tasks. Large business processes are just too complex to be jointly expressed with all data objects in a single model. However, mobile apps with small-scale tasks and processes are well suited to this kind of integrated modelling approach. The evaluation has shown that users appreciate the simplicity of the editor without switching between multiple interrelated models, a major distinction from related approaches to graphical mobile app development. Possibly related to the aforementioned modelling support, not even programmers miss the two-step approach of first specifying a global data model and then separately defining the respective processes. Nevertheless, as potential future extension, an optional view of the inferred data model may be interesting for them to check the modelling result before generation. Similarly, two non-technical users stated the wish for a preview of the resulting screens. However, both suggestions are neither meant to be editable nor mandatory for the app creation process and rather serve as reassuring validation while modelling the use case. It can therefore be said that modelling activities should suit the users' previous experience, potentially ignoring established concepts of technical domains for the greater good of a more comprehensible and seamless modelling environment.

As a result, bringing mobile app modelling to this new level of abstraction not only bridges the gap to the field of business process modelling but can also impact organizations. On the one hand, new technical possibilities arise from process-centric app models. For example, already documented business processes can be used as input for cross-platform development targeting a variety of heterogeneous mobile devices. On the other hand, codeless app generation creates the opportunity for different development methodologies. The distinction between app developer and framework developer can lead to performance benefits and better resource utilization on hardware-constrained devices such as smartphones. Best practices of mobile software development can be adopted by developers with expert knowledge of the respective platforms within the transformations which are then applied consistently throughout all generated apps. It has been shown that structural implementation decisions and even small-scale code refactorings can significantly improve battery consumption and execution times [52,53]. Also, instead of involving domain experts only in requirements phases before the actual development, an equitable relationship with fast development cycles is possible because changes to the model can be deployed instantly. Furthermore, future non-technical users may themselves develop applications according to their needs, extending the idea of self-service IT to its actual development. All of these ideas, however, rely on the modelling support provided by the environment, as begun with MAML's data model inference mechanism. Smart software to guide and validate the created models is required instead of simply representing the digital equivalent of a sheet of paper. In the future, graphical editors may evolve beyond just organizing and linking different models, towards tools enabling novel digital ecosystems through supportive technology.

7 Conclusion

In this work, a model-driven approach to mobile app development called MAML was presented which focuses around a declarative and platform-agnostic DSL to graphically create mobile business apps. The visual editor component provides advanced modelling support such as suggestions and validation through automatic data model inference. In addition, transformations allow for a codeless generation of app source code for multiple platforms. To evaluate the notation with regard to comprehensibility and usability, an extensive observational study with 26 participants was performed. The results confirm the design goals of achieving a wide-spread comprehensibility of MAML models for different audiences of software developers, process modellers, and domain experts. In comparison to the IFML notation, an equivalent MAML model is perceived as much less complex – in particular by non-technical users – and participants felt a high level of control, thus confidently solving their tasks. Furthermore, we analysed the challenges when extending the cross-platform approach to multiple app-enabled device classes. The applicability of MAML for this so-called *pluri-platform development* was assessed using a second study on a newly developed generator for the Wear OS smartwatch platform. As a result, MAML's approach of describing a mobile app as process-oriented set of use cases reaches a suitable balance between the technical intricacies of cross-platform app development and the simplicity of usage through the high level of abstraction and can be used to create app source code for both device classes from the same input models.

In case of the presented study results, some limitations may threaten their validity. Although a reasonable amount of participants was chosen for the observational interviews, additional evaluations may be carried out after the next iteration of MAML's development. Also, our participants were mostly students which potentially reduces the generalizability of the results. However, their generation of app-experienced adults already participates in the general workforce and can be seen as realistic (albeit not representative) sample. The synthetic examples within the case study were designed to test a wide range of MAML's capabilities and uncover usability issues. Therefore, a real-world application would strengthen the validity of the approach and at the same time represents future work.

Regarding limitations of the approach itself, the chosen level of abstraction requires assumptions on the generic representation of data in the prototype. Possibilities to customize low-level details such as UI styling for different device classes need to be addressed in future, for example on the level of the intermediate MD^2 representation. Also, improvements of the generator prototype itself are part of ongoing work to provide a wide set of platform-adapted representations.

The presented process-oriented DSL offers the opportunity for research on a suitable framework structure for pluri-platform development and possible reuse of common transformations among multiple generators. Also, the process of developing such a framework of coupled components through a team with different roles may be investigated to further integrate model-driven techniques with traditional software development. Technically, further iterations on the framework's development are planned in order to provide additional user support, improve performance, and incorporate feedback based on the observed usability issues. Finally, the applicability of our approach to create business apps through model-driven transformations of MAML's platform-agnostic models to further device classes with drastically different UIs such as smart virtual assistants also presents exciting possibilities for future research.

References

1. van der Aalst, W.: Formalization and verification of event-driven process chains. Inf. Softw. Technol. **41**(10), 639–650 (1999). https://doi.org/10.1016/S0950-5849(99)00016-6
2. Alulema, D., Iribarne, L., Criado, J.: A DSL for the development of heterogeneous applications. In: FiCloudW, pp. 251–257 (2017)
3. Apache Software Foundation: Apache Cordova documentation (2019). https://cordova.apache.org/docs/en/latest/
4. Apple Inc: watchOS (2019). www.apple.com/watchos/
5. Architecture Board ORMSC: Model driven architecture (MDA): Document number ormsc/2001-07-01 (2001). http://www.omg.org/cgi-bin/doc?ormsc/2001-07-01
6. Bangor, A., Kortum, P., Miller, J.: Determining what individual SUS scores mean: adding an adjective rating scale. J Usability Stud. **4**(3), 114–123 (2009)
7. Bargh, J.A., Chartrand, T.L.: Studying the mind in the middle: a practical guide to priming and automaticity research. In: Judd, C.M., Reis, H.T. (eds.) Handbook of Research Methods in Social and Personality Psychology, pp. 253–285. Cambridge University Press, New York (2000)

8. Barnett, S., Avazpour, I., Vasa, R., Grundy, J.: A multi-view framework for generating mobile apps. In: IEEE Symposium on Visual Languages and Human-Centric Computing (VL/HCC), pp. 305–306 (2015). https://doi.org/10.1109/VLHCC.2015.7357239
9. Bizness Apps: Mobile app maker—bizness apps (2019). http://biznessapps.com/
10. Brambilla, M., Dosmi, M., Fraternali, P.: Model-driven engineering of service orchestrations. In: 5th World Congress on Services (2009). https://doi.org/10.1109/SERVICES-I.2009.94
11. Breu, R., Kuntzmann-Combelles, A., Felderer, M.: New perspectives on software quality [guest editors' introduction]. IEEE Softw. **31**(1), 32–38 (2014). https://doi.org/10.1109/MS.2014.9
12. Brooke, J.: SUS-a quick and dirty usability scale. In: Jordan, P.W., Thomas, B., Weerdmeester, B.A., McClelland, A.L. (eds.) Usability Evaluation in Industry, pp. 189–194. Taylor and Francis, London (1996)
13. Bubble Group: Bubble - visual programming (2019). https://www.bubble.is/
14. Buchmann, T.: Bxtend - a framework for (bidirectional) incremental model transformations. In: 6th International Conference on Model-Driven Engineering and Software Development (MODELSWARD) (2018). https://doi.org/10.5220/0006563503360345
15. da Silva, L.P., Brito e Abreu, F.: Model-driven GUI generation and navigation for android BIS apps. In: 2014 2nd International Conference on Model-Driven Engineering and Software Development (MODELSWARD), pp. 400–407 (2014)
16. El-Kassas, W.S., Abdullah, B.A., Yousef, A.H., Wahba, A.M.: Taxonomy of cross-platform mobile applications development approaches. Ain Shams Eng. J. (2015). https://doi.org/10.1016/j.asej.2015.08.004
17. Esakia, A., Niu, S., McCrickard, D.S.: Augmenting undergraduate computer science education with programmable smartwatches. In: SIGCSE, pp. 66–71 (2015). https://doi.org/10.1145/2676723.2677285
18. France, R.B., Ghosh, S., Dinh-Trong, T., Solberg, A.: Model-driven development using UML 2.0: promises and pitfalls. Computer **39**(2), 59–66 (2006). https://doi.org/10.1109/MC.2006.65
19. Fuller, J.B., Hester, K., Barnett, T., Frey, L., Relyea, C., Beu, D.: Perceived external prestige and internal respect: new insights into the organizational identification process. Hum. Relat. **59**(6), 815–846 (2006). https://doi.org/10.1177/0018726706067148
20. Gamma, E., Helm, R., Johnson, R., Vlissides, J.: Design Patterns: Elements of Reusable Object-Oriented Software. Addison-Wesley Professional Computing Series. Addison-Wesley, Reading (1995)
21. Gediga, G., Hamborg, K.C.: Evaluation in der software-ergonomie. J. Psychol. **210**(1), 40–57 (2002). https://doi.org/10.1026//0044-3409.210.1.40
22. GoodBarber: Goodbarber: Make an app (2019). https://www.goodbarber.com/
23. Google Inc: J2ObjC (2018). http://j2objc.org/
24. Google Inc: Wear OS by Google smartwatches (2019). https://wearos.google.com/
25. Granada, D., Vara, J.M., Brambilla, M., Bollati, V., Marcos, E.: Analysing the cognitive effectiveness of the WebML visual notation. Softw. Syst. Model. (2015). https://doi.org/10.1007/s10270-014-0447-8
26. Hemel, Z., Visser, E.: Declaratively programming the mobile web with Mobl. In: Conference on Object Oriented Programming Systems Languages and Applications (OOPSLA), pp. 695–712. ACM (2011). https://doi.org/10.1145/2048066.2048121
27. International Organization for Standardization: ISO 5807:1985 (1985)

28. International Organization for Standardization: ISO 9241-110:2006 (2006)
29. Jesdabodi, C., Maalej, W.: Understanding usage states on mobile devices. In: ACM International Joint Conference on Pervasive and Ubiquitous Computing, UbiComp, pp. 1221–1225. ACM (2015). https://doi.org/10.1145/2750858.2805837
30. Jones, C., Jia, X.: The AXIOM model framework: transforming requirements to native code for cross-platform mobile applications. In: 2nd International Conference on Model-Driven Engineering and Software Development (MODELSWARD). IEEE (2014)
31. Koren, I., Klamma, R.: The Direwolf inside you: end user development for heterogeneous web of things appliances. In: Bozzon, A., Cudre-Maroux, P., Pautasso, C. (eds.) ICWE 2016. LNCS, vol. 9671, pp. 484–491. Springer, Cham (2016). https://doi.org/10.1007/978-3-319-38791-8_35
32. Langlois, B., Jitia, C.E., Jouenne, E.: DSL classification. In: The 7th OOPSLA Workshop on Domain-Specific Modeling (2007)
33. Ludei Inc: Canvas+ Cocoon documentation (2019). https://docs.cocoon.io/article/canvas-engine/
34. Majchrzak, T.A., Ernsting, J.: Reengineering an approach to model-driven development of business apps. In: Wrycza, S. (ed.) SIGSAND/PLAIS 2015. LNBIP, vol. 232, pp. 15–31. Springer, Cham (2015). https://doi.org/10.1007/978-3-319-24366-5_2
35. Majchrzak, T.A., Ernsting, J., Kuchen, H.: Achieving business practicability of model-driven cross-platform apps. OJIS **2**(2), 3–14 (2015)
36. Majchrzak, T.A., Wolf, S., Abbassi, P.: Comparing the capabilities of mobile platforms for business app development. In: Wrycza, S. (ed.) SIGSAND/PLAIS 2015. LNBIP, vol. 232, pp. 70–88. Springer, Cham (2015). https://doi.org/10.1007/978-3-319-24366-5_6
37. Mernik, M., Heering, J., Sloane, A.M.: When and how to develop domain-specific languages. ACM Comput. Surv. **37**(4), 316–344 (2005). https://doi.org/10.1145/1118890.1118892
38. Moody, D.: The "physics" of notations: towards a scientific basis for constructing visual notations in software engineering. IEEE Trans. Softw. Eng. **35**(5), 756–778 (2009)
39. Neate, T., Jones, M., Evans, M.: Cross-device media: a review of second screening and multi-device television. Pers. Ubiquitous Comput. **21**(2), 391–405 (2017). https://doi.org/10.1007/s00779-017-1016-2
40. Object Management Group: Business process model and notation (2011). http://www.omg.org/spec/BPMN/2.0
41. Object Management Group: Unified modeling language (2015). http://www.omg.org/spec/UML/2.5
42. Pentaho Corp: Data integration - kettle (2017). http://community.pentaho.com/projects/data-integration/
43. Product Hunt: 7 tools to help you build an app without writing code (2016). https://medium.com/product-hunt/7-tools-to-help-you-build-an-app-without-writing-code-cb4eb8cfe394
44. Reiter, A., Zefferer, T.: Power: a cloud-based mobile augmentation approach for web- and cross-platform applications. In: CloudNet, pp. 226–231. IEEE (2015). https://doi.org/10.1109/CloudNet.2015.7335313
45. Rieger, C.: Business apps with MAML: a model-driven approach to process-oriented mobile app development. In: Proceedings of the 32nd Annual ACM Symposium on Applied Computing, pp. 1599–1606 (2017)

46. Rieger, C.: Evaluating a graphical model-driven approach to codeless business app development. In: 51st Hawaii International Conference on System Sciences (HICSS), pp. 5725–5734 (2018)
47. Rieger, C.: MAML code respository (2019). https://github.com/wwu-pi/maml
48. Rieger, C., Kuchen, H.: A process-oriented modeling approach for graphical development of mobile business apps. Comput. Lang. Syst. Struct. **53**, 43–58 (2018). https://doi.org/10.1016/j.cl.2018.01.001
49. Rieger, C., Kuchen, H.: A model-driven cross-platform app development process for heterogeneous device classes. In: 52nd Hawaii International Conference on System Sciences (HICSS), pp. 7431–7440 (2019)
50. Rieger, C., Majchrzak, T.A.: A taxonomy for app-enabled devices: mastering the mobile device jungle. In: Majchrzak, T.A., Traverso, P., Krempels, K.-H., Monfort, V. (eds.) WEBIST 2017. LNBIP, vol. 322, pp. 202–220. Springer, Cham (2018). https://doi.org/10.1007/978-3-319-93527-0_10
51. Rivera, J., van der Meulen, R.: Gartner says by 2018, more than 50 percent of users will use a tablet or smartphone first for all online activities (2014). http://www.gartner.com/newsroom/id/2939217
52. Rodriguez, A., Mateos, C., Zunino, A.: Improving scientific application execution on android mobile devices via code refactorings. Softw. Pract. Exp. **47**(5), 763–796 (2017). https://doi.org/10.1002/spe.2419
53. Sahar, H., Bangash, A.A., Beg, M.O.: Towards energy aware object-oriented development of android applications. Sustain. Comput. Inform. Syst. **21**, 28–46 (2019). https://doi.org/10.1016/j.suscom.2018.10.005
54. Simons, C., Wirtz, G.: Modeling context in mobile distributed systems with the UML. J. Vis. Lang. Comput. **18**(4), 420–439 (2007). https://doi.org/10.1016/j.jvlc.2007.07.001
55. Singh, K., Buford, J.: Developing WebRTC-based team apps with a cross-platform mobile framework. In: IEEE CCNC (2016). https://doi.org/10.1109/CCNC.2016.7444762
56. Spool, J., Schroeder, W.: Testing web sites: five users is nowhere near enough. In: CHI 2001 Extended Abstracts on Human Factors in Computing Systems, pp. 285–286. ACM (2001). https://doi.org/10.1145/634067.634236
57. The Eclipse Foundation: Sirius (2019). https://eclipse.org/sirius/
58. Thomson, G.: BYOD: enabling the chaos. Netw. Secur. **2012**(2), 5–8 (2012). https://doi.org/10.1016/S1353-4858(12)70013-2
59. Umuhoza, E., Brambilla, M.: Model driven development approaches for mobile applications: a survey. In: Younas, M., Awan, I., Kryvinska, N., Strauss, C., Thanh, D. (eds.) MobiWIS 2016. LNCS, vol. 9847, pp. 93–107. Springer, Cham (2016). https://doi.org/10.1007/978-3-319-44215-0_8
60. Vaupel, S., Taentzer, G., Harries, J.P., Stroh, R., Gerlach, R., Guckert, M.: Model-driven development of mobile applications allowing role-driven variants. In: Dingel, J., Schulte, W., Ramos, I., Abrahão, S., Insfran, E. (eds.) MODELS 2014. LNCS, vol. 8767, pp. 1–17. Springer, Cham (2014). https://doi.org/10.1007/978-3-319-11653-2_1
61. Virzi, R.A.: Refining the test phase of usability evaluation: how many subjects is enough? Hum. Factors **34**(4), 457–468 (1992)
62. WebRatio: WebRatio (2019). http://www.webratio.com

63. Wolber, D.: App inventor and real-world motivation. In: 42nd ACM Technical Symposium on Computer Science Education (SIGCSE) (2011). https://doi.org/10.1145/1953163.1953329
64. Xamarin Inc: Developer center - Xamarin (2019). https://developer.xamarin.com
65. Zyła, K.: Perspectives of simplified graphical domain-specific languages as communication tools in developing mobile systems for reporting life-threatening situations. Stud. Log. Gramm. Rhetor. **43**(1) (2015). https://doi.org/10.1515/slgr-2015-0048

What Matters for Chatbots? Analyzing Quality Measures for Facebook Messenger's 100 Most Popular Chatbots

Juanan Pereira[✉] and Óscar Díaz

ONEKIN Research Group, University of the Basque Country (UPV/EHU),
Bilbao, Spain
{juanan.pereira,oscar.diaz}@ehu.eus

Abstract. Chatbots are becoming mainstream. This work aims at ascertaining what are the enablers behind this popularity. To this end, we introduce four quality attributes, namely, "support of a minimal set of commands", "foresee language variations", "human-assistance provision" and "timeliness". These criteria are applied to the 100 most popular Facebook Messenger chatbots. We review and measure both capacities and performance in order to find correlations between quality attribute fulfilment and popularity (chatbots' 'likes'). Results show no significance correlations between quality attributes and chatbot popularity. However, the experiment comes up with three main contributions. First, a detailed description of how to measure these four quality attributes. Second, insights about how this assessment can be automatized, paving the way towards chatbot-evaluation platforms. Third, a checklist of frequently committed interaction errors as found in the revised chatbots. This might help developers to double-check their development.

Keywords: Conversational agents · Chatbots · Mobile UI · Messaging

1 Introduction

A chatbot is a computer program which conducts a conversation via instant messaging. Chatbots are becoming pervasive: from flight reservation to purchase tracking, from math quizzing to University entry advice, chatbots are percolating throughout an increasing number of activities. Chatbots have been around in online web based environments for quite some time. But the breakthrough comes from chatbots moving to a new realm: instant messaging. Benefits are numerous. They 'live' in a familiar chat interface and builds upon existing infrastructure of mobile and social commerce (e.g. ability to initiate a conversation or send

© Springer Nature Switzerland AG 2019
T. A. Majchrzak et al. (Eds.): *Towards Integrated Web, Mobile,
and IoT Technology*, LNBIP 347, pp. 67–82, 2019.
https://doi.org/10.1007/978-3-030-28430-5_4

interesting updates[1], a common known-interface[2]). Downloading and installing apps is no longer necessary, and the use of smart-phones allows for easily accessing/monitoring personal data [9]. Furthermore, the use of chatbots can be more cost effective than human-assisted support [7]. On top, chatbots are platform independent as they use the messenger infrastructure. This makes app downloading redundant.

These benefits explain the staggering growing figures exhibited by chatbots both in terms of users, savings and satisfaction. According to Gartner [4], by 2020, over 50% of medium to large enterprises will have deployed product chatbots. As for satisfaction, 38% of consumers rated their overall perception of bots as positive [1], in comparison with only 11% who rate their interaction as negative. Nevertheless, not all chatbots are born equal. Even for the same platform, chatbots enjoy a wide range of popularity as the number of "likes" so confirms. This begs the question of what are those "likes" linked to.

Certainly, natural-language capabilities make a difference w.r.t. previous technologies. And much work has been devoted to this issue in the academic literature: [17] investigates methods to train and adapt a bot to a specific user's language use via a user-supplied training corpus; [16] rises the issue of the difficulty in evaluating of domain-oriented chatbots as for natural language understanding and reasoning; [6] studies chatbots' conversational abilities and context sensitiveness (i.e. dialogue context detection, coherent dialogues, ability to repair a dialog when parameters are missing, rich vocabulary). These studies illustrate the preponderant role given to natural language abilities in the academic literature. However, the grey literature (i.e. blogs, wikis and the like) sustains a different angle where aspects other than natural-language understanding might be equally important for chatbot success [15]. Here, practitioners value chatbot technology not so much for their human-like capabilities but their help to deliver results [15]. This does not mean that there is not value in making chatbots more human-like. This certainly plays a key role in user engagement/comprehension but other factors could be equally important. This work attempts to apprehend those other factors. Hence, this work tackles the following research question: **is there any correlation between these non-natural-language quality attributes and chatbot success?**

The answer to this question is important for new developments to find a balance between the new features (e.g., machine learning, natural-language abilities) and the more traditional features (e.g., conversational interfaces) in order to optimize developers' efforts. We aim at ascertaining whether a correlation exists between traditional features and the chatbot popularity. Specifically:

– we take "*likes*" as the measure of chatbot popularity,

---

[1] Mobile-bots have the push-message ability, updating users with interesting news whenever they happen.

[2] Web-bots were implemented ad-hoc, no standard UI was available for them to use, in contrast with mobile-bots running inside well-know apps, like Messenger, Telegram, Skype...

- we introduce four quality attributes: "support of a minimal set of commands", "foresee language variations", "human-assistance provision" and "timeliness". The selection is grounded on scholarly but also reputed blog references from 2016 and 2017. In addition, we only use attributes that could be semi-automatically extracted from interacting with the bot via a script, getting rid of any subjectivity (Sect. 3),
- we analyze whether there is any correlation between the number of *likes* and these quality attributes for the top 100 most popular chatbots at Facebook Messenger (Sect. 4)
- as a by-product, common shortcomings found among the 100 analyzed chatbots are also listed (Sect. 5)

This article extends our work [13] presented at SAC'18 (Web Technologies Track), by adding a detailed explanation of how to develop the scripts for measuring chatbot quality features.

2   Background

Wikipedia defines a chatbot as "a computer program which conducts a conversation via auditory or textual methods. Such programs are often designed to convincingly simulate how a human would behave as a conversational partner, thereby passing the Turing test". This definition rests on the first examples of chatbots back in 1966 where they were programmed to respond to a user's questions with simple matching patterns. Today, they possess sophisticated techniques to understand users' questions and deliver useful and relevant responses, but the stress is still in the NLP side. Some authors question this stress: "Early attempts at chatbots have fallen flat in their execution, mostly because they have relied too much on natural language processing or A.I. capabilities that simply don't yet exist" [3]. Other authors highlight the role of chatbots as "a combination of multiple services as it can combine communication services with information services, entertainment and commercial transactions" [19], or their role as a personal assistant capable of providing a range of services [18].

This change in perception is fueled by a change in technology. In contrast with first-generation chatbots, mobile chatbots outperform previous technologies in different ways:

- offering a sense of ubiquity. Mobile chatbots are always available
- broader knowledge base. Current chatbots can answer questions about many knowledge-areas
- ability to be pro-active, i.e. initiating a conversation or sending interesting updates. Mobile bots have the push-message ability, updating users with interesting news whenever they happen,
- providing a common known-interface. Web-bots were implemented ad-hoc, no standard UI was available for them to use, in contrast with mobile-bots running inside well-know apps, like Messenger, Telegram, Skype...

- ability to integrate with third-party APIs (like Uber, Google Flights, Deliveroo, etc.)
- wider multimedia elements. Mobile chatbots capitalize smart-phone wealth I/O resorts. Image, audio and video can be captured to send rich-media messages to the chatbot and the other way around: chatbots can answer using the same media elements
- geo-position minded. This feature allows to program geo-context dependent bots, opening a whole new set of context dependent applications. Some examples follow: offers in specific geo-areas, commuting schedule suggestions, deep information about events happening near the user, ...

This boils down to chatbot developers struggling to keep up with a myriad of technologies in constant flux. This begs the question of how to balance the new features (machine learning, natural language processing abilities) with more traditional chatbot features (conversational interfaces) in order to optimize developers' efforts. This in turn, advices to look into the extent traditional features impact chatbot success. This moves us to the next section.

3   Quality Attributes for Chatbots

For Web-based chatbots, the most recent evaluation study we are aware of is the one by [6]. Here, bots are evaluated along the visual look (e.g. cartoon-like animation, living-person video), embedded mechanism (e.g. floating window, pull-out side bar, etc.), speech synthesis unit (e.g. availability of a speech synthesis unit), knowledge base (e.g. answering capabilities for questions involving knowledge at the level of an elementary school graduate), availability of clickable links that trigger behavior in a context sensitive way, personalization options (e.g. change of the visualization gender, access to the conversation history, recalling the user name, etc.), and emergency response in unexpected situations (e.g. typos, misspellings and various errors commonly appear in users' statements). On the other hand, Radziwill et al. conducts a literature review on 32 papers, summarizing efficiency, effectiveness and user satisfaction chatbot quality attributes [14]. Many of those characteristics tend to be subjective and hence, difficult to be subject to automatization. By contrast, we are interested in analyzing a large number of chatbots in order to find correlations. This requires quality attributes to be obtained through scripting, and most importantly, being able to be replicated. Therefore, this Section not only introduces quality attributes but a way for these attributes to be automatically worked out.

Support of a Minimal Set of Commands. Bots should exhibit a common set of frequently used commands, e.g. "Help" (to ask for help), "Hi" (for saluting), or "Cancel" (for canceling the current conversation flow). It is likely users might have already used other bots. Tapping on a shared vocabulary for common scenarios will certainly reduce the learning bar. Unfortunately, bot designer are not always aware of this fact. Figure 1 shows the case where "Help" does not lead to a description of how to interact with the bot, i.e. the expected behaviour (we asked for help and the bot answered with a simple greeting). Similar situations happens for "Cancel" where some bots insist on using "Start over" or "Menu", to start over a conversation.

How to Measure. Our tests specifically checks for the existence of the *Help* and the *Cancel* commands. These are not mandatory, but our experience (and final results so confirms) they are common practice.

Fig. 1. Counter example: When asking for help, bots should describe what commands do they offer (unlike the bot in the picture)

Foresee Language Variations in Command Input. Bots don't need to have a top-notch natural language understanding to succeed. That said, a minimum variation in their commands is recommended. Although it will be impossible to avoid confusing dialects and other non-traditional forms of communication, the bot can hone in on the consumers' needs by offering hints. For example, avoid open-ended questions and offer multiple-choice answers [11]. The expectations are for bots to handle at least those commands that are available via button. The presence of buttons do not make their text counterparts redundant. Users might stick to the text, and directly type the commands. This double-channel interface is not always planned. Figure 2 displays a case in point. Here, the user types "Daily Round-Up". The bot does not recognize this interaction despite being one of the button-supported offers. Click on the "Daily Round-Up" button and the bot correctly detects our intention.

How to Measure. Our testing script starts by clicking on a button. Next, the button label is typed as text, and the outputs are compared. Finally, typing errors (plural forms, spelling errors, ...) and synonyms are tried out, and their outcome compared. Interesting enough, even simple variations are not always accounted for. Figure 3 shows the case for the `mancity` bot, where typing "Hi" instead of "Hello" makes the bot crumble.

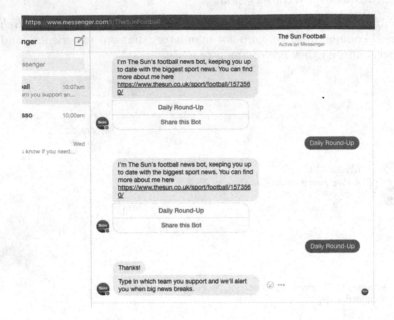

Fig. 2. Counter example: Bots should respond to text messages that mimic the same text of button label (unlike the bot in the figure)

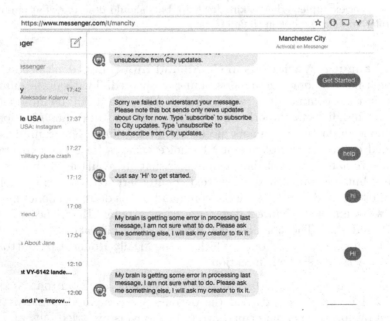

Fig. 3. Counter example: Not handling basic language variation: "Hi" for "Hello"

Exhibit Language Variations in Command Output. In search for humanity, chabots might add small, random variations: not answering always exactly with the same limited set of words.

How to Measure. We used one of the commands available twice, and checked the answers. They should be almost the same but, in best case scenario, exhibit minor differences (using different emoticons, for example).

Provide Human Assistance. Do not try and trick your audience into thinking they are speaking with a human rather than a bot [11]. Bots could not be prepared to answer all questions related to their businesses, and the possibility to contact human assistance should be there [3]. On the other end of the spectrum, some companies just create a chatbot carcass whose only purpose is to transition from Messenger to the traditional call-center. It rests to be seen whether this strategy pays off.

How to Measure. Our script resorts to some heuristics to detect this scenario. Variations on the sentence "I'd like to talk to a person" are worked out ("to the manager", "to an operator") as well as checking for keywords (e.g. "redirect", "contact", "human"). Nevertheless, some cases require manual checking to confirm the redirection.

Efficiency. Efficiency is always a matter when the user is waiting. As reported in Sect. 4.2, most of the reviewed bots answer in less than 2 s. However some bots have a much greater delay, even to the extent of having to wait for days or weeks for an answer. Obviously, this kind of bots are just simple re-directors to humans. However, they don't specifically state that, leaving users wondering what they should do.

How to Measure. We measure the answer time for the "Help" command, twice for each bot, obtaining mean values (in milliseconds).

Table 1 summarizes the described measures and cites literature references that back the metric.

Table 1. Bot metrics to measure.

Metric	Evidence backing sources	Metric Tag
NL Expressiveness vs. buttom-style dialog	[6, 10, 15, 19]	M1
Usage of emojis	[12]	M2
Answer variations	[2]	M2
Presence of the *help* command	Authors of this article	M3
Presence of the *cancel* command	Authors of this article	M3
Typos resilience	[5]	M4
Availability of human redirection	Authors of this article	M5
Answer Delay	Authors of this article	M6

4   Experiment

About the Platform. This section describes how aforementioned quality attributes have been measured for the 100 most popular chatbots in Facebook Messenger. This platform enjoys over 1.2 billion users, with more than 60 million companies sending messages every day (reaching 2 billion messages) [20]. The platform accounts for over 30,000 bots [8] with over 150,000 developers registered by April 2017.

Facebook offers an API for bot programming[3]. However, it lacks any facilities to test those bots. In fact, some third-party frameworks attempt to overcome this gap by simulating the Facebok API. Anyway, those frameworks are meant for testing your own bot, rather than someone else's bot.

About the Chats Under Study. We obtained a list of the 100 chatbots that rank top on *likes* according to the *chatbottle* directory (https://chatbottle.co).

Table 2. An excerpt of the 25-first bots' metrics values: "✓" and "x" stands for passing or failing the metric, respectively. Metrics include: presence of the *cancel* command (*CANCEL*); support for language variations in input (*NLP* and *TYPO*); language variations in command output (VARIANCE); support for human assistance (*HUMAN*); extent of delays in bots' answers (DELAY)

id	botname	Likes	NLP (M1)	VARIANCE (M2)	CANCEL (M3)	TYPO (M4)	HUMAN (M5)	DELAY (M6)
1	Maroon5	39265054	x	✓	✓	x	x	2250
2	50cent	37267790	x	x	x	x	x	1200
3	NBA	32828370	x	x	x	x	x	2176
4	victoriassecret	28004345	x	x	✓	x	✓	1042
5	cnn	27147415	x	x	x	x	x	1233
6	samsungmobileusa	25556849	✓	✓	x	x	✓	1246
7	mancity	24334688	x	x	x	x	x	2782
8	Burberry	17229903	x	x	x	x	✓	942
9	dominos	17199558	x	x	✓	x	x	1420
10	goal	16166262	x	x	✓	x	x	1399
11	tommyhilfiger	11053612	✓	x	x	✓	✓	3659
12	Jwoww	9653608	x	x	x	x	x	1026
13	al-jazeera	9306060	✓	x	✓	✓	x	1698
14	skyscanner	9038401	✓	x	x	✓	✓	1911
15	djhardwell	8671283	x	x	x	x	✓	1323
16	TheWeatherChannel	7688430	✓	x	x	x	x	6992
17	theguardian	7290368	x	x	x	x	✓	1520
18	peoplemag	6803648	✓	x	✓	x	x	1480
19	WSL	6074759	✓	x	x	✓	x	1800
20	wsj	5614978	x	x	x	x	x	2041
21	robbiewilliams	4983836	x	✓	x	✓	✓	1193
22	christinamilian	4691331	x	x	✓	x	✓	1004
23	wholefoods	4163177	✓	x	✓	✓	✓	972
24	redfoo	3902041	x	x	✓	x	x	1300
25	mtvnews	3842979	x	x	x	✓	x	2787

---

[3] https://developers.facebook.com/docs/messenger-platform.

Likes is the number of people that liked the bot in Facebook. It can be considered a direct and plausible evidence of success that can be easily checked, though this number could be artificially inflated by bot owners. Using a browser automation library (*Puppeteer*), we initiated a web conversation with each one. Analyzing the HTML code of the initial message, we extracted each bot's numerical identifier and number of *likes*. This bot ID opens the door to interact with the bot programmatically.

About the Procedure. Our evaluation framework rests on a custom NodeJS solution that leverages *Puppeteer*, a library for managing a Chromium headless browser[4]. Our scripts are publicly available in Github[5] for any stakeholder interested. In addition, manual supervision was required to face captcha requests. Section 4.1 describes main points of the implementation details.

Problems. Facebook's abuse protection measures include: (1) banning of users accessing Messenger from a cloud server IP address, and (2) banning users for exceeding the maximum number of new connections with bots for each user and day (this measures are thought to protect spamming the platform). Once a user triggers any of these alerts, first Facebook will show a captcha that the user should manually solve. If the user insists on override the limits, they will be banned, temporally or, for recurrent offenders, for good.

4.1 Implementation Details

Our solution resorts to Puppeteer, a Node library which provides a high-level API to control the headless Chrome browser over the DevTools protocol. Using Puppeteer, our implemented analyzer communicates with Facebook Messenger bots programmatically, simulating real-user actions.

The analyzer can be executed as a standalone command line script or as the back-end of a Web application to offer real-time graphical updates to the user, transmitting the information via webSockets.

The following sections define the inner workings of the solution, describing the main classes and sequence diagrams of both the analyzer and Web front-end main functions.

There are three main classes (*Analyzer, HelperMessages, HelperPuppeteer*) and three wrapper classes (*WebInterface, Server, ScriptLauncher*) involved in the solution (Fig. 4).

Analyzer: this is the main class, contains the function responsible for performing the complete analysis of the chatbot. This includes: obtaining number of likes, Messenger identificator, percentage of emojis and multimedia elements used, the time it takes for the bot to respond to each message, its ability to process the text commands (as if the specific button had been pressed), the usefulness of the initial and help messages, support of variations in the answers and finally, support for understanding messages with small typos.

---

[4] https://github.com/GoogleChrome/puppeteer.
[5] https://github.com/petuscov/puppeteerFacebook.

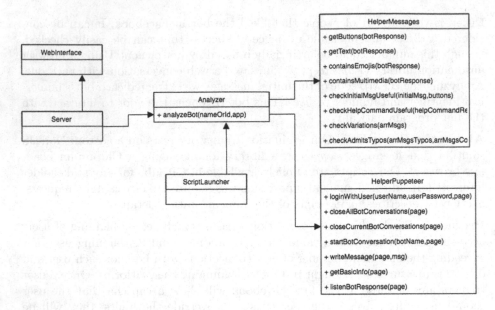

Fig. 4. Class diagram of the solution

HelperPuppeteer: this class performs more complex interactions with Puppeteer, facilitating actions such as logging in Messenger, closing conversations, initiating them with the indicated bots, and sending and receiving messages.

HelperMessages: a class composed of utility functions, it is used to extract information from the object that represents the response of a bot, as it is processed by the corresponding function of the class mentioned above, HelperPuppeteer.

ScriptLauncher: a simple class that acts as a wrapper of the chatbots analysis function, to perform bots analysis directly from the command-line console.

Server: it also acts as a wrapper and calls the analysis function, but in addition, it also opens the Web interface and bridges it with the analysis in progress, updating the web front-end data obtained from the bot in real time.

Figure 5 shows the sequence diagram of the *listenBotResponse* function, which is responsible for returning an object with the components of a bot response in Messenger. This function is part of the *helperPuppeteer NodeJS* module, and makes use, in turn, of *eval*, a function of the Puppeteer library which allows the execution of code in headless Chrome. For detecting changes (bot answers) in the DOM of the web page that we are interacting with, we leverage the MutationObserver API.

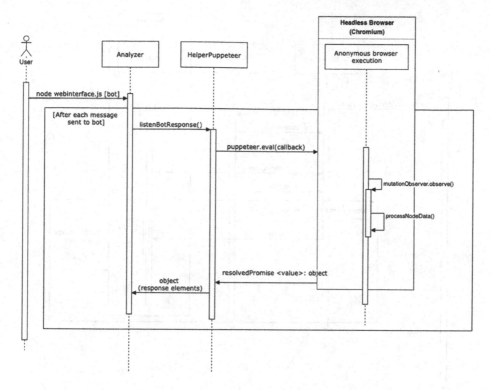

Fig. 5. Sequence diagram of the listener (listenBotResponse) method

The *eval* function includes as its parameters three helper functions necessary for processing the bot response: *processNodeData*, *getBottomButtons* and *getPath*. Respectively, processNodeData analyzes the content of the received messages, and *getBottomButtons* detects if buttons are displayed at the bottom of the conversation. Both functions make use of *getPath*, since if it is the case that there are lower buttons or if the buttons are displayed in a message, it is necessary to obtain the path of these to be able to press them through a Puppeteer action. *processNodeData* is also responsible of extracting the text of the messages, and detect whether they contain images, multimedia elements or emojis.

Figure 6 shows the sequence diagram that represents the flow that takes place when launching the analysis with a web front-end. Running the script, a server is launched and calls the analysis function, which in turn, communicates with the server through webSockets (connection and update methods in the diagram). The server subsequently opens a browser connecting with the web front-end. Once this has been done, the successive information updates of the analyzer are notified to the server, and this is transferred to the web interface (Fig. 7 shows a module of the web interface, where a graph depicts the time elapsed for successive 'goodbye' messages).

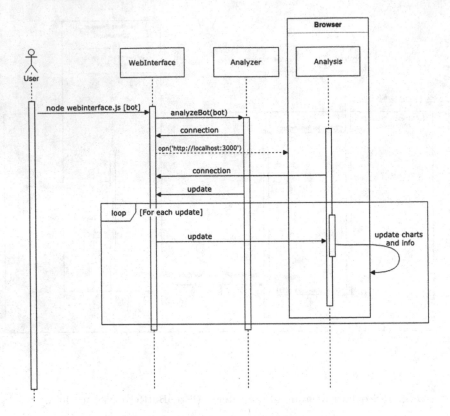

Fig. 6. Sequence diagram of the web interface method

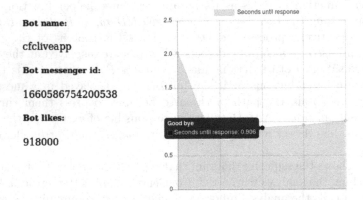

Fig. 7. The web front-end, showing a real-time updated graph with response delays and basic metadata of a bot.

4.2 Results

Table 2 summarizes the results obtained for the first 25 bots. Delays greatly vary among chatbots due to the fact that some of them are simple re-directors to humans. If we take apart those cases, and focus on real bots, the minimum delay is of 697 ms, maximum of 6992 ms, with a mean of 1975 ms and a standard deviation of 1376 ms. That is, on average, a near 2 s delay which seems acceptable.

With respect to basic NLP support, only 30% of the most popular 100 list support basic natural-language recognition features and 24% of them are also able to recognize commands with small typos. More surprising is the low number of chatbots that support the `Cancel` command (38%). In some case, the `Help` command t left a great deal to be desired (see Fig. 1). With regard to adding random pieces of text to the output, only 14% of the bots use this approach to simulate answer entropy. In the upside, 46% of the bots account for redirection to human beings.

4.3 Correlation Analysis

We wanted to know if there exist any kind of correlation between the metrics and the number of *likes* the chatbot receives.

Using R, we obtain a linear regression model with all the factors of Table 2 using this simple formula:

```
model <- lm(likes~NLP+variance+cancel+typo+human,data=data)

Coefficients:
            Estimate Std. Error t value Pr(>|t|)
(Intercept) 10933207    2668229   4.098 0.000172 ***
NLPv         -171069    3873724  -0.044 0.964971
variancev   -4999246    4273076  -1.170 0.248184
cancelv     -1865723    3169168  -0.589 0.558999
typov       -2877588    4201346  -0.685 0.496907
humanv      -2810310    3074076  -0.914 0.365486
```

As we can see, none of the correlation terms is significant. The variance explained by the coefficients is also very low. In fact, if we had followed those coefficients, the popularity of the bot would have been inversely proportional to implementing features like NLP or answer variance. The data also tells us a metric for acceptable answering delays: 2 s.

5 Discussion

First results seem to support the insight that most popular IM chatbots don't even try to impersonate a real human conversation. They are quite mechanic in their answers, and lack basic interaction and communication patterns. Many of the chatbots are quite mechanic in their behavior, just showing a button-based navigation interface and lacking the minimum skills of text based commands recognition. Their popularity is definitively not related to the degree of their language or human skills, but, more probably, it is related to the already existing popularity of the brand that the chatbot represents. Bots under the name of Manchester City, CNN, NBA or any other popular brand will automatically receive a lot of public attention. Only a few chatbots from the 100 most popular ones are really connected to indie or no-mass-media brands.

Interacting with the most popular Messenger bots also help to detect some common programming errors. Next subsection reports on this experience in terms of a checklist for developers to quickly check them out.

5.1   Checklist of Frequently Observed Issues in Messenger Chatbots

Welcome message

- Does it fit in the screen of a smart-phone?
- Have you taken into account i18n issues? (instead of "Get Started", what happens when receiving a Spanish counterpart "Empezar" message?)
- Do you show a selection of your commands list alongside your first message?

Minimal set of common commands

- Do you support a "Cancel" command? (that aborts the last conversation orand redirects the user to a starting menustate)
- Do you support a "Help" command? (that shows some available commands)

Basic language variations support

- Do you support text messages that are equivalent to the text commands enclosed in buttons? (e.g. if your bot offers two buttons with Yes/No options, do you accept also "Yes" and "No" typed answers?)
- Do you use any library or technique to allow small typos in your users' messages?
- Do you support basic conversation recovery techniques? (e.g. when it stumbles upon an unknown command?)

Variance

- Do you add some variations to your text answers? (rephrase your answers, add emojis, add gifs, human expressions...)

Human redirection

- Do you have an option to contact a human being?

Efficiency

- Do you answer in a few seconds or is your bot a simple to-human re-director? If the latter, do you inform the user about your average response time? is e-mail an option?

6   Conclusions

We analyzed the 100 most popular bots of Facebook Messenger against four quality attributes, namely: "support of a minimal set of commands", "foresee language variations", "human-assistance provision" and "timeliness". The aim: looking for correlations between these attributes and chatbot popularity in terms of number of "likes". Results show that there is no significance correlation with any of these attributes. Technically, we also offer a framework for researchers to replicate the outcomes of the article, or adapt them to fit other bot platforms. Developers and maintainers could leverage this platform to test their bots against a series of research-backed, desirable bot features and metrics.

These findings lead us to conjecture that chatbots popularity is basically influenced by the fame of the brand behind, or the usefulness of the information obtained from the bot. This seems to be akin to Gartner's survey where 48% of users preferred a bot that solves their issues rather than a bot with "personality" [4]. Nevertheless, two interesting insights can be drawn:

- newcomers do not need to exclusively focus on NLP or in performing human-like dialogues in order to achieve a popular bot. Obviously this conclusions should not be interpreted as an invitation to avoid those techniques, but as a motivation for new developers to dare to program a bot even though they lack skills on NLP or conversational interfaces,
- bot programming seems to be still in its infancy. This seems to be suggested by the fact that only 12 out of 100 are programmed to avoid repetition in their responses, a clear sign of being a bot rather than a human. Likewise, studied chatbots frequently resort to humans when the conversation gets stuck, rather than conducting the user to other questions to overcome this scenario. We appreciate this situation in 40 chatbots (e.g. darngoodyarn,1-800-flowers, Opla, LawTrades). Finally, the lack of common communication patterns prevent users from move their experiences and expectations from one chatbot to another.

This situation will most certainly change in the short run as a new crop of chatbot platforms are facilitating performant chatbot development. This should go hand-in-hand with quality-testing frameworks, and evidence-based quality attributes.

References

1. Beaver, L.: Chatbots are gaining traction (2017). https://www.businessinsider.com/chatbots-are-gaining-traction-2017-5. Business Insider
2. Coniam, D.: The linguistic accuracy of chatbots: usability from an ESL perspective. Text Talk **34**(5), 545–567 (2014)
3. Emmet Connolly: Principles of bot design (2016). https://blog.intercom.com/principles-bot-design/. Inside Intercom
4. Gartner: Chatbots will appeal to Modern Workers (2018). https://www.gartner.com/smarterwithgartner/chatbots-will-appeal-to-modern-workers/

5. Klüwer, T.: From chatbots to dialog systems. Conversational agents and natural language interaction: techniques and effective practices, pp. 1–22 (2011)
6. Kuligowska, K.: Commercial chatbot: performance evaluation, usability metrics and quality standards of embodied conversational agents. Prof. Cent. Bus. Res. **2**, 1–16 (2015)
7. Lester, J., Branting, K., Mott, B.: Conversational agents. In: The Practical Handbook of Internet Computing, pp. 220–240 (2004)
8. de Looper, C.: Facebook is now letting you pay for things through the Messenger Platform, September 2016. https://www.digitaltrends.com/social-media/facebook-messenger-chatbot-payments/
9. Messina, C.: 2016 will be the year of conversational commerce, January 2016. https://medium.com/chris-messina/2016-will-be-the-year-of-conversational-commerce-1586e85e3991
10. Morrissey, K., Kirakowski, J.: 'Realness' in chatbots: establishing quantifiable criteria. In: Kurosu, M. (ed.) HCI 2013. LNCS, vol. 8007, pp. 87–96. Springer, Heidelberg (2013). https://doi.org/10.1007/978-3-642-39330-3_10
11. Newlands, M.: Top 10 Practices for Making A Great Chatbot, June 2017. https://www.entrepreneur.com/article/296358
12. Pauletto, S., et al.: Exploring expressivity and emotion with artificial voice and speech technologies. Logop. Phoniatr. Vocology **38**(3), 115–125 (2013)
13. Pereira, J., Díaz, Ó.: A quality analysis of Facebook Messenger's most popular chatbots. In: Proceedings of the 33rd ACM/SIGAPP Symposium on Applied Computing, pp. 33:2144–2150 (2018)
14. Radziwill, N.M., Benton, M.C.: Evaluating Quality of Chatbots and Intelligent Conversational Agents. arXiv preprint arXiv:1704.04579 (2017)
15. Ramos, R.: Screw the Turing test - chatbots don't need to act human (2017). https://venturebeat.com/2017/02/03/screw-the-turing-test-chatbots-dont-need-to-act-human/
16. Goh, O.S., Ardil, C., Wong, W., Fung, C.C.: A black-box approach for response quality evaluation of conversational agent systems. Int. J. Comput. Intell. Syst. **3**, 37–41 (2012)
17. Shawar, B.A., Atwell, E.: Different measurements metrics to evaluate a chatbot system. In: Proceedings of the Workshop on Bridging the Gap: Academic and Industrial Research in Dialog Technologies, pp. 89–96, April 2007
18. Sullivan, L.: Facebook Chatbots Hit 70% Failure Rate As Consumers Warm Up To The Tech (2017). https://www.mediapost.com/publications/article/295718/facebook-chatbots-hit-70-failure-rate-as-consumer.html
19. Van Eeuwen, M., Van Der Kaap, H.: Mobile conversational commerce: messenger chatbots as the next interface between businesses and consumers (2017). http://essay.utwente.nl/71706/1/vanEeuwen_MA_BMS.pdf
20. Wiggers, K.: Facebook Messenger can now find bots, order food, and more (2017). https://www.digitaltrends.com/social-media/facebook-messenger-news-f8-2017/

Linguistic Abstractions for Interoperability of IoT Platforms

Maurizio Gabbrielli[1], Saverio Giallorenzo[2(✉)], Ivan Lanese[1],
and Stefano Pio Zingaro[1]

[1] Università di Bologna / INRIA, Bologna, Italy
{maurizio.gabbrielli,stefanopio.zingaro}@unibo.it, ivan.lanese@gmail.com
[2] University of Southern Denmark, Odense, Denmark
saverio.giallorenzo@gmail.com

Abstract. The Internet of Things (IoT) advocates for multi-layered platforms—from edge devices to Cloud nodes—where each layer adopts its own communication standards (media and data formats). While this freedom is optimal for in-layer communication, it puzzles cross-layer integration due to incompatibilities among standards. Also enforcing a unique communication stack within the same IoT platform is not a solution, as it leads to the current phenomenon of "IoT islands", where disparate platforms hardly interact with each other. In this paper we tackle the problem of IoT cross-layer and cross-platform integration following a language-based approach. We build on the Jolie programming language, which provides uniform linguistic abstractions to exploit heterogeneous communication stacks, allowing the programmer to specify in a declarative way the desired stack, and to easily change it, even at runtime. Jolie currently supports the main technologies from Service-Oriented Computing, such as TCP/IP, Bluetooth, and RMI at transport level, and HTTP and SOAP at application level. We integrate in Jolie the two most adopted protocols for IoT communication, i.e., CoAP and MQTT. We report our experience on a case study on Cloud-based home automation, and we present high-level concepts valuable both for the general implementation of interoperable systems and for the development of other language-based solutions.

1 Introduction

The Internet of Things (IoT) advocates for multi-layered software platforms, each adopting its own media protocols and data formats [1–3]. The problem of integrating layers of the same IoT platform, as well as different IoT vertical solutions, involves many levels of the communication stack, spanning from link-layer communication technologies, such as ZigBee and WiFi, to application-layer protocols like HTTP, CoAP [4,5], and MQTT [6,7], reaching the top-most layers of data-format integration [8].

Technology-wise, architects of IoT platforms can choose between two approaches at odds. The first approach favors optimal in-layer communications,

© Springer Nature Switzerland AG 2019
T. A. Majchrzak et al. (Eds.): *Towards Integrated Web, Mobile,*
and IoT Technology, LNBIP 347, pp. 83–114, 2019.
https://doi.org/10.1007/978-3-030-28430-5_5

i.e., choosing media protocols and data formats best suited for the interactions happening among homogeneous elements, e.g., edge devices (connectionless protocols and binary data formats [3]), mid-tier controllers (gateways and aggregators on the RESTful stack [9]), or Cloud nodes (scalable publish-subscribe message queues [10]). Following this first approach is optimal for in-layer communication. However, at the cross-layer level, the heterogeneity and possible incompatibility of the chosen standards make enforcing integrity within the IoT system complex and the resulting integration fragile. The second architectural approach favors cross-layer consistency, enforcing a unique communication stack over a single IoT platform. Here cross-layer integration is simpler thanks to the adoption of a single medium and data format. However such enforced uniformity is the main cause of the phenomenon known as "IoT island" [11,12], where IoT platforms take the shape of vertical solutions that provide little support for collaboration and integration with each other. How to overcome this limitation is currently a hot topic, tackled also by ongoing EU projects, e.g., symbIoTe [12] and bIoTope [13].

In this paper we tackle the problem of IoT integration (both cross-layer and cross-platform) following a language-based approach focused on integration at both the transport (TCP or UDP) and application layer. To reach our goal we do not start from scratch, but we leverage the work done in the area of Service-Oriented Architectures (SOAs) [14] and we build on the Jolie programming language [15–18]. In particular, we rely on those abstractions provided by Jolie that (*i*) let different communication protocols seamlessly coexist and interoperate within the same program and (*ii*) let programmers dynamically choose which communication stack should be used for any given communication. Concretely, we fork the Jolie interpreter—written in Java—into a prototype called JIoT [19], standing for "Jolie for IoT". JIoT supports all the protocols already supported by the Jolie interpreter (TCP at the transport level, and protocols such as SOAP, RMI and HTTP at the application level), while adding the application-level protocols for IoT, namely CoAP (and, as a consequence, UDP at the transport level) and MQTT. Notably, when the application protocol supports different representation formats (such as JSON, XML, etc.) of the message payload, as in the case of HTTP and CoAP, JIoT, like Jolie, can automatically marshal and un-marshal data as required.

We structure our presentation as follows. We overview in Sect. 2 our approach and summarize our contribution in Sect. 3. Then, we discuss the main challenges we faced in our development in Sect. 4, we present how a programmer can use CoAP/UDP (Sect. 5) and MQTT (Sect. 6) in JIoT, and we detail our implementation in Sect. 7. We describe, in Sect. 8, a scenario on Cloud-based home automation where a JIoT architecture coordinates heterogeneous edge devices. Finally, we position our contribution with respect to related work in Sect. 9 and we draw final remarks in Sect. 10.

JIoT is available at [19], released under the LGPL v2.1 license. The code snippets reported in this paper are based on version 1.2 of JIoT. The integration of JIoT into the official code-base of the Jolie language is ongoing work.

2   Approach Overview

Without proper language abstractions, guaranteeing interoperability among protocols belonging to different technology stacks is highly complex. The problem is further exacerbated when one has to modify the technology stack used for some specific interaction. The replacement may be either static, e.g., because of the deployment of new, heterogeneous devices in a pre-existing system, or dynamic, e.g., to support a changing topology of disparate mobile devices. Contrarily, with JIoT most of the complexity of guaranteeing interoperability is managed by the language interpreter and hidden from the programmer.

As an illustrative example of the proposed approach, let us consider a scenario where we want to integrate two islands of IoT devices, both collecting temperature data, but relying on different communication stacks, namely HTTP over TCP and CoAP over UDP. The end goal is to program a collector which receives and aggregates temperature measurements from both islands.

Following the structure of Jolie programs, the collector programmed in JIoT is composed of two parts: a *behavior*, specifying the logic of the elaboration, and a *deployment*, describing in a declarative way how communication is performed. This separation of concerns is fundamental to let programmers easily change which communication stack to use, preserving the same logic for the elaboration.

As an example of program behavior, let us consider the code below, where main is the entry point of execution of Jolie programs.

```
1 main {
2   ...
3   receiveTemperature( data );
4   ...
5 }
```

Above, line 3 contains a reception statement. Receptions in Jolie indicate a point where the program waits to receive a message. In this case, the collector waits to receive a temperature measurement on *operation* receiveTemperature (an operation in Jolie is an abstraction for technology-specific concepts such as channels, resources, URLs, ...). Upon reception, it stores the retrieved value in variable data. Besides the logic of computation of the collector, we also need to specify the deployment, i.e., on which technologies the communication happens; in the example above, how the collector receives messages from other devices. In Jolie this information is defined within *ports*. For example, the port to receive (denoted with keyword inputPort) HTTP measurements can be defined as in Listing 1. Port CollectorPort1 specifies that the collector expects inbound communications via Protocol http using a TCP/IP socket receiving at URL "localhost" on TCP port 8000. A port exposes a set of operations, collected within a set of Interfaces. In the example, the input port CollectorPort1 declares to expose interface TemperatureInterface, which is defined at lines 1–3 of Listing 1. The interface declares the operation receiveTemperature, including the type of expected data (string), as a OneWay operation, namely an asynchronous communication that does not require any reply from the collector (except the acknowledgment automatically provided by the TCP implementation).

```
1 interface TemperatureInterface {
2  OneWay: receiveTemperature( string )
3 }
4
5 inputPort CollectorPort1 {
6  Location: "socket://localhost:8000"
7  Protocol: http
8  Interfaces: TemperatureInterface
9 }
```

Listing 1. Example of interface and input port in Jolie.

Thanks to port `CollectorPort1`, the collector can receive data from the HTTP island. To integrate the second island, we just need to define an additional port, similar to `CollectorPort1`, except for using UDP/IP datagrams at the transport layer and CoAP [4,5] at the application layer. Hence, the whole code of the collector becomes:

```
1 interface TemperatureInterface {
2  OneWay: receiveTemperature( string )
3 }
4
5 inputPort CollectorPort1 {
6  Location: "socket://localhost:8000"
7  Protocol: http
8  Interfaces: TemperatureInterface
9 }
10
11 inputPort CollectorPort2 {
12  Location: "datagram://localhost:5683"
13  Protocol: coap
14  Interfaces: TemperatureInterface
15 }
16
17 main {
18  ...
19  receiveTemperature( data );
20  ...
21 }
```

Listing 2. Code of the Collector Example.

The example above highlights how, using the proposed language abstractions, the programmer can write a unique behavior and exploit it to receive data sent over heterogeneous technology stacks. Indeed, the `receiveTemperature` operation takes measurements from both the `inputPorts`. For instance, if communication over `CollectorPort2` fails, port `CollectorPort1` can still receive data. Programmers can also specify elaborations that depend on the used technologies, by using different operations in different ports. Jolie supports both inbound and outbound communications, the latter declared with outputPorts, whose structure follows that of inputPorts. Furthermore, the Location and Protocol of outputPorts can be changed at runtime, enabling the dynamic selection of the appropriate technologies for each context.

As mentioned, Jolie enforces a strict separation of concerns between behavior, describing the logic of the application, and deployment, describing the communication capabilities. The behavior is defined using the typical constructs of structured sequential programming, communication primitives, and operators to deal with concurrency (parallel composition and input choices [17]).

Jolie communication primitives comprise two modalities of interaction.

Outbound OneWay communications send a message asynchronously, while RequestResponse communications send a message and wait for a reply (they capture the well-known pattern of request-response interactions [20]). Dually, inbound OneWay communications wait to receive a message, without sending a reply, while inbound RequestResponses wait for a message and send back a reply.

Jolie supports many communication media (via keyword Location) and data protocols (via keyword Protocol) in a simple, uniform way. This is one of the main features of the Jolie language, and the reason why we base our approach on it. Each communication port declares the medium and data protocol used to communicate, hence, to switch to a different technology stack, one just needs to change the declaration of Location and Protocol of a given port. As expected, the behavior (i.e., the actual logic of computation) of any Jolie program is unaffected by any change to its ports. Hence, a Jolie program can provide the same service (i.e., the same behavior) through different media and protocols just by specifying different deployments. Being born in the field of SOAs, Jolie supports the main technologies from that area: e.g., communication media like TCP/IP sockets, Bluetooth L2CAP, Java RMI, and Unix local sockets; and data protocols like HTTP, JSON-RPC, XML-RPC, SOAP and their respective SSL versions.

3   Contribution

To substantiate the effectiveness of our language-based approach to IoT integration, we add to Jolie support for the main communication stacks used in the IoT setting. Concretely, the added contribution of JIoT with respect to Jolie is the integration of two application protocols relevant in the IoT scenario, namely CoAP [4,5] and MQTT [6,7]. Notably, in JIoT the usage of such protocols is supported by the same linguistic abstractions that Jolie uses for SOA protocols such as HTTP and SOAP.

Even if Jolie provides support for the integration of new protocols, when set in the context of IoT technology, the task is non trivial. Indeed, all the protocols previously supported by Jolie exploit the same internal interface, based on two assumptions: (i) the usage of underlying technologies that ensure reliable communications and (ii) a point-to-point communication pattern.

However, those assumptions do not hold when considering the two IoT technologies we integrate:

- CoAP communications can be unreliable since they are based on UDP connectionless datagrams. CoAP provides options for reliable communications, however these are usually disabled in an IoT setting, since it is important to preserve battery and bandwidth;
- MQTT communications are based on the publish-subscribe paradigm, which contrasts with the point-to-point paradigm underlying the Jolie communication primitives. Hence, we need to define a mapping to express publish-subscribe operations in terms of Jolie communication abstractions. In doing so, we need to balance two factors: (i) preserving the simplicity of use of the point-to-point communication style and (ii) capturing the typical publish-subscribe flow of communications. Such a mapping is particularly challenging in the case of request-response communications. Remarkably, the mapping that we present in this work is general and could be used also in other contexts.

This paper integrates, revises, and extends material from [21], where we presented, discussed, and provided basic technical details on the proposed language-based approach to IoT integration. Main extensions comprise:

- advanced technical details on our implementation (Sect. 7) including:
  - a general account on how media and protocols are separated from the Jolie interpreter and how they can be developed as independent modules;
  - extensive details on the implementation of UDP, CoAP, and MQTT protocols;
- a comprehensive case study on a home automation scenario (Sect. 8) where we consider:
  - local, cross-layer communication among things and mid-tier controllers (edge devices and fog nodes);
  - remote, cross-layer interactions among Cloud nodes and mid-tier controllers.

We conclude this section briefly discussing the current limitations of JIoT related to its usage in the programming of low-level edge devices—like Arduinos and other microcontrollers. JIoT supports dynamic scenarios where the nodes in the network can switch among many technology stacks according to internal or environmental conditions, such as available energy or quality of communication. From preliminary discussions with collaborators and IoT practitioners, we collected positive opinions on the idea of using JIoT for programming low-level edge devices. Given these positive remarks, we investigated the feasibility of running JIoT programs over edge devices, possibly including additional language abstractions to provide low-level access to in-board sensors and actuators. However, our survey revealed a market of devices fragmented over incompatible hardware architectures and characterized by strong constraints over both computational power and energy consumption. Considering these limitations, we concluded that supporting the execution of JIoT-like programs over edge devices would require a strong engineering effort. While this research direction is promising, we deem it non-urgent, since currently developers tend to program very simple behaviors for edge devices [3], which usually capture some data (e.g., through one of their sensors) and then send them to mid-to-top-tier devices. The latter usually process and coordinate the flow of data: they have powerful hardware, they communicate over reliable channels, and they have fewer (if any) constraints with respect to battery/energy consumption.

Considered the discussion above, in this work we omit the low-level programming of edge devices and we focus on mid-to-top-tier ones, which can host the JIoT runtime and which, given their topological context, directly benefit from the flexibility of the approach.

4   JIoT: Jolie for IoT

Jolie currently supports some of the main technologies used in SOAs (e.g., HTTP, SOAP). However, only a limited amount of IoT devices uses the media and protocols already supported by Jolie. Indeed, protocols such as CoAP [4,5] and MQTT [6,7], which are widely used in IoT scenarios, are not implemented in Jolie. Integrating these protocols, as we have done, is essential to allow Jolie programs to directly interact with the majority of IoT devices. We note that emerging frameworks for interoperability, such as the Web of Things [22], rely on the same protocols we mentioned for IoT, thus JIoT is also compliant with them. However there are some challenges linked to the integration of these technologies within Jolie:

- *lossless vs. lossy protocols* — In SOAs, machine-to-machine communication relies on lossless protocols: there are no strict constraints on energy consumption or bandwidth and it is not critical how many transport-layer messages are needed to ensure reliable delivery. That is not true in IoT networks, where communication is constrained by energy consumption, which defines what technology stack can be used. Indeed, many IoT communication technologies, among which the mostly renowned CoAP application protocol, rely on the UDP transport protocol — a connectionless protocol that gives no guarantee on the delivery of messages, but allows one to limit message exchanges and, by extension, energy and bandwidth consumption. Since Jolie assumes lossless communications, the inclusion of connectionless protocols in the language requires careful handling to prevent misbehaviors;
- *point-to-point vs. publish-subscribe* — The premise of the Jolie language is to provide communication constructs that do not depend on a specific technology. To do so, the language assumes a point-to-point communication abstraction, which is common to many protocols like HTTP and CoAP. However, to integrate the MQTT protocol in Jolie, we need to model Jolie point-to-point semantics as MQTT publish-subscribe operations.

Indeed, Jolie already provides language constructs usable with many communication protocols, hence the less disruptive approach is to use the same constructs, which are designed for a point-to-point setting, also for MQTT. This requires to find for each point-to-point construct a corresponding effect in the publish-subscribe paradigm. The final result is that the execution of a given Jolie behavior is similar under both point-to-point and publish-subscribe technologies.

5   Supporting Constrained Application Protocol in Jolie

The *Constrained Application Protocol* (CoAP) [4,5] is a specialized web transfer protocol for constrained scenarios where nodes have low power and networks are lossy. The goal of CoAP is to import the widely adopted model of REST architectures [23] into an IoT setting, that is, optimizing it for machine-to-machine applications. In particular, like HTTP, CoAP makes use of GET, PUT, POST, and DELETE methods. Following the RFC [5], CoAP is implemented on top of the UDP transport protocol [24], with optional reliability. Indeed, CoAP provides two communication modalities: a reliable one, obtained by marking the message type as confirmable (CON), and an unreliable one, obtained by marking the message type as non confirmable (NON).

As an example, we consider a scenario with a controller, programmed in JIoT, that communicates with one of many thermostats in a home automation scenario. Thermostats are accessible at the generic address `"coap://localhost/##"` where `"##"` is a two-digit number representing the identifier of a specific device. Each thermostat accepts two kinds of interactions: a GET request on URI `"coap://localhost/##/getTemperature"`, that returns the current temperature, and a POST request on URI `"coap://localhost/##/setTemperature"`, that sets the temperature of the HVAC (heating, ventilation, and air conditioning) system.

We comment below Listing 3, where we report the code of a possible JIoT controller that interacts with a specific thermostat.

Our scenario includes two CoAP resources, referred to as `"/getTemperature"` and `"/setTemperature"`. We model them in JIoT at lines 4–7 of Listing 3, by defining the `interface ThermostatInterface`, which includes a `RequestResponse` operation `getTmp`, representing resource `"/getTemperature"`, and a `OneWay` operation `setTmp`, representing resource `"/setTemperature"`. By default, we map operation names to resource names, hence in our example we would need resources named `"/getTmp"` and `"/setTmp"`, respectively. However one can override this default by defining the coupling of resource names and operations as desired. This allows programmers to use interfaces as high level abstractions for interactions, while the grounding to the specific case is done in the deployment. Here we purposefully choose to use operation names that differ from resource names to underline that the two concepts are related but loosely coupled. On the one hand the coupling between the name of the resource and the operation can be seen as a way of quickly binding actions exposed by the CoAP server with operations.

```
1  type getTmpType: void { .id: string }
2  type setTmpType: int { .id: string }
3
4  interface ThermostatInterface {
5   RequestResponse: getTmp( getTmpType )( int )
6   OneWay: setTmp( setTmpType )
7  }
8
9  outputPort Thermostat {
10  Location: "datagram://localhost:5683"
11  Protocol: coap {
12   .osc.getTmp << {
13    .messageCode = "GET",
14    .contentFormat = "text/plain",
15    .messageType = "CON",
16    .alias = "/%!{id}/getTemperature"
17   };
18   .osc.setTmp << {
19    .messageCode = "POST",
20    .messageType = "CON",
21    .alias = "/%!{id}/setTemperature"
22   }
23  }
24  Interfaces: ThermostatInterface
25 }
26
27 main {
28  getTmp@Thermostat( { .id = "42" } )( temp );
29  if ( temp > 27 ) {
30   setTmp@Thermostat( 24 { .id = "42" } )
31  } else if ( temp < 15 ) {
32   setTmp@Thermostat( 22 { .id = "42" } )
33  }
34 }
```

Listing 3. JIoT controller communicating over CoAP/UDP.

On the other hand decoupling resource names and operations permits to handle more complex deployments where, for instance, a single operation responds for different resources. At lines 9–25 we define an outputPort to interact with the Thermostat. At line 10 we specify the Location of the thermostat. Recalling that the scheme of the resources of the thermostats is "coap://localhost/##/...", we define the Location of the port using the UDP "datagram://" protocol, followed by the first part of the resource schema "localhost" and the UDP port on which it accepts requests. Here we assume thermostats to use CoAP standard UDP port, which is "5683". Note that, in the Location, we do not define the address of a specific thermostat, e.g., "datagram://localhost:5683/42". On the contrary, we just specify the generic address to access thermostats in the system, while the specific binding will be done at runtime, thanks to the .alias parameter of the coap protocol, described later on.

At line 11 we define coap to be the protocol used by the outputPort. At lines 12–22 we specify some parameters of the coap protocol — this matches the standard way in which Jolie defines parameters for Protocols in ports. Here, we follow the methodology presented in [25] for the implementation of the HTTP protocol in Jolie — indeed CoAP adopts HTTP naming schema and resource interaction methods. In particular, we draw from [25] the parameter prefix .osc, whose name is the acronym of "operation-specific configuration" and which is used for configuration parameters related to a specific operation.

In the example, we define .osc parameters for both operations getTmp and setTmp. At line 13 we specify that the CoAP verb used for operation getTmp is "GET". At line 14 we define, using the .contentFormat parameter, that the encoding of the payload of the message is in text format. Other accepted values for the .contentFormat parameter are "json" and "xml". Marshalling and un-marshalling is automatic and transparent to the programmer. This feature is enabled by the structure of Jolie variables, which are always tree-shaped, hence they can be easily translated into representations based on that shape. At line 15 we set the .messageType parameter to "CON", that stands for confirmable. Accepted values for the .messageType parameter are confirmable and not confirmable ("NON"), the latter being the default value. In the first case the sender will receive an acknowledgment message from the receiver, in the second case it will not. At line 16, following the practice introduced in [25], we specify that getTmp aliases a resource whose path concatenates a static part, given by the Location, and the instantiation of the template "/%!{id}/getTemperature" provided by protocol parameter .alias. The template is instantiated using values from the parameter of the operation invocation in the behavior, e.g., value 42 at line 28[1]. Hence, the interpretation of the declaration at line 16 is that, when invoking operation getTmp at runtime, the element id of the invocation will be removed from the payload and used to form the address of the requested resource. The aliasing for operation setTmp (line 21) is similar to that of getTmp, while the operation uses verb POST. Since here the .contentFormat parameter is omitted, the default "text/plain" is used.

To conclude, we briefly comment the runtime execution of the example, described in the behavior at lines 28–33. At line 28 the controller invokes operation getTmp. Being an outgoing RequestResponse, the invocation defines on which port to perform the request (Thermostat) and presents two pairs of round brackets: the first contains the data for the request, the second points to the variable that will store the received response. Recalling the aliasing defined at line 16, at line 28 we define the value of element id = 42, thus the URI of the resource invoked at runtime is "coap://localhost/42/getTemperature". Notably, in the example we hard-coded the id of the device, however in a more realistic setting the value of id would be retrieved dynamically, e.g., as an execu-

---

[1] In Jolie the dot . defines path traversals inside trees. Hence, the notation {.id = 42} indicates a tree with an empty root and a subnode called id, whose value is 42.

tion parameter, from a configuration file or from a database. Once received, the response from thermostat 42 is assigned to variable temp. The example concludes with a conditional in which, if the temperature is above 27° (line 29), the thermostat is set to lower room temperature to 24°, while, if the temperature lies below 15°, the thermostat is set to raise the temperature to 22°.

Dually to outputPorts, inputPorts allow the programmer to specify inbound communications. The parameters described above are valid also for inputPorts, with the only difference that messageType works only for RequestResponses, and specifies whether the communication of the reply is reliable or not. Note that, concerning the .alias parameter, the template is instantiated using the address of the incoming communication and the values are inserted among the elements of the payload.

6 Supporting Message Queue Telemetry Transport in Jolie

Message Queue Telemetry Transport (MQTT) [6,7] is a publish/subscribe messaging application protocol built on top of the TCP transport protocol.

A typical publish/subscribe interaction pattern can be diagrammatically represented as in Fig. 1 where:

1. a Subscriber subscribes to topic (a) at some Broker;
2. a Publisher publishes a message to topic (a) at the same Broker;
3. the Broker forwards the message to topic (a) to the Subscriber.

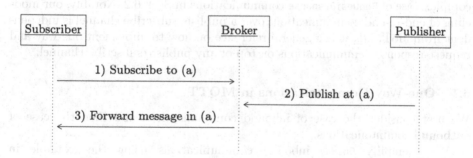

Fig. 1. Typical publish/subscribe interaction pattern.

More generally, messages published on a topic are forwarded to all current subscribers for the topic.

On top of the basic mechanism of publish/subscribe, MQTT defines three levels of quality of service (QoS) for the delivery of each message published by a publisher. QoS levels determine whether messages can be lost and/or duplicated. Concretely, QoS levels are as follows:

– *At most once* — the message can be lost, no duplication can occur.

```
1 interface TemperatureInterface {
2  OneWay: receiveTemperature( string )
3 }
4
5 inputPort CollectorPort3 {
6  Location: "socket://localhost:8050"
7  Protocol: mqtt {
8   .broker = "socket://localhost:1883"
9  }
10  Interfaces: TemperatureInterface
11 }
12
13 main {
14  ...
15  receiveTemperature( data );
16  ...
17 }
```

Listing 4. Code of the Collector Example, revised for MQTT.

– *At least once* — delivery of the message is guaranteed, but duplication may occur.
– *Exactly once* — delivery of the message is guaranteed and duplication cannot occur.

To present how we model the MQTT protocol in JIoT, we first detail the simpler case of OneWay communications in Sect. 6.1. Then, we address the more complex case of RequestResponse communications in Sect. 6.2. Notably, our modeling of end-to-end communications over a publish/subscribe channel is independent from JIoT, i.e., it is a general reference on how to implement one-way and request-response communications on top of any publish/subscribe channel.

6.1   One-Way Communications in MQTT

We first consider the case of inbound communications and then the case of outbound communications.

We exemplify OneWay inbound communications using the example in Listing 4, which is a revision of the example in Listing 2 by omitting the ports CollectorPort1 and CollectorPort2 and by adding an MQTT inputPort named CollectorPort3.

As expected, the program behavior and the structure of the inputPort are unchanged. Main novelties are:

– the used Location (line 6) has the prefix "socket://" (as seen in the HTTP port) since MQTT relies on TCP transport protocol;
– the used Protocol (line 7) is mqtt;
– the .broker protocol parameter (line 8), which is compulsory when the mqtt protocol is used in inputPorts, specifies the address of the Broker.

From the perspective of the programmer, the syntax and the effect of the communication primitive are the same as in Listing 2. However, we actually exchange several messages to capture that effect in MQTT, as shown in Fig. 2.

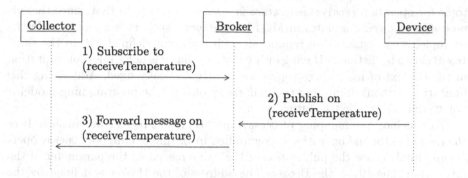

Fig. 2. Representation of the example in Listing 4.

Beyond defining such message exchanges, we also need to decide how to identify the topic on which the message exchange is performed.

Regarding the message exchanges, from the point of view of the programmer, an inbound OneWay communication receives a datum from the communication partner. To obtain the same effect using the publish/subscribe paradigm, one has first to subscribe at the Broker to the chosen topic and then wait to receive a message on that topic, forwarded by the Broker. How topics are selected will be detailed later on. The execution of a reception on a OneWay operation comprises two actual communications: a subscription from the program to the Broker and a message delivery in the opposite direction. However, subscription to topics and the execution of a message reception are logically separated and can be done at different moments. Indeed, the subscription is performed when the JIoT program is launched for all operations present in MQTT inputPorts. This choice is more in line with the expected behavior of Jolie programs — and of Service-Oriented programs in general — where messages to operations whose reception statements are not yet enabled are stored until the actual execution of the reception. Here, if the subscription is performed along with the execution of the OneWay operation, previous messages could be no more available. In JIoT, the compulsory parameter .broker is needed precisely to know the address at which the subscription needs to be performed. The address for the delivery of the actual message is the usual Location of the inputPort.

Regarding the selection of topics, similarly to what done for CoAP resources, in MQTT by default we map JIoT operations to topics, otherwise we use the .osc parameter .alias to loose the coupling between operations and topics. We remark that alias parameters in inputPorts have a different behavior in MQTT with respect to HTTP and CoAP. In CoAP the name of the resource extracted from the received message is used to derive the instantiation of the .alias template. The values resulting from the match are then inserted among the elements of the payload before storing it in the target variable data. Instead, in MQTT, the .alias parameter is used to identify the topic for subscription. For example, in Listing 4, one could add the Protocol parameter .osc.receiveTemperature.alias = "temperature" to specify that the selected topic for operation receiveTemperature is "temperature". Note that, since there is no outgoing data, templates in MQTT inputPorts, such as "temperature" in the example, are constants (we require all such constants defined within the same inputPort to be distinct). Having only constant aliases is not a relevant limitation in the context of IoT, where topics are mostly statically fixed. Addressing this limitation without disrupting the uniformity of the Jolie programming model is not trivial and it is left as future work.

To conclude the mapping of OneWay operations in MQTT, we consider here the case of outbound operations, exemplified in Listing 5. Outgoing OneWay operations simply cause the publication of the value passed as the parameter of the invocation (line 19) at the Broker. The address of the Broker is defined by the Location (line 6) of the outputPort Broker. The topic is derived from the name of the operation and the parameter of the invocation, using protocol parameter .alias as usual. Being an MQTT publication, we specify the .QoS protocol parameter (line 10), which selects the QoS level "Exactly once" for the operation setTmp. Similarly to what we have done in CoAP with the contentFormat protocol

```
1  interface ThermostatInterface {
2   OneWay: setTmp( TmpType )
3  }
4
5  outputPort Broker {
6   Location: "socket://localhost:1883"
7   Protocol: mqtt {
8    .osc.setTmp << {
9     .format = "raw",
10    .QoS = 2, // exactly once QoS
11    .alias = "%!{id}/setTemperature"
12   }
13  }
14  Interfaces: ThermostatInterface
15 }
16
17 main {
18  ...
19  setTmp@Broker( 24 { .id = "42" } );
20  ...
21 }
```

Listing 5. Example of outgoing MQTT OneWay communication.

parameter, we define in .format the encoding of the message payload, in this case a "raw" stream of bytes.

```
1  interface ThermostatInterface {
2   RequestResponse: getTmp( TmpType )( int )
3  }
4
5  outputPort Broker {
6   Location: "socket://localhost:1883"
7   Protocol: mqtt {
8    .osc.getTmp << {
9     .format = "raw",
10    .QoS = 2, // exactly once QoS
11    .alias = "%!{id}/getTemperature",
12    .aliasResponse = "%!{id}/getTempReply"
13   }
14  }
15  Interfaces: ThermostatInterface
16 }
17
18 main {
19  ...
20  getTmp@Broker( { .id = "42" } )( temp );
21  ...
22 }
```

Listing 6. JIoT controller communicating over MQTT.

6.2   Request-Response Communications in MQTT

To discuss RequestResponse communications, let us consider the example in Listing 3, revised in Listing 6 by replacing the CoAP protocol with MQTT. We omit OneWay communications and concentrate on the outbound RequestResponse. Afterwards, we will also discuss the dual inbound RequestResponse.

Syntactically, the main novelty with respect to the outputPort in Listing 5 is the addition of Protocol parameter .aliasResponse. This parameter specifies the name of the topic where the receiver will publish its response.

From the point of view of the programmer, an outbound RequestResponse is composed of an outgoing communication followed by an inbound reply. The outgoing communication is implemented using the approach already seen for OneWay communications, i.e., using the .alias Protocol parameter to identify the topic. Then, one has the issue of relating the outgoing request with its reply. Many standard point-to-point communication technologies, such as HTTP/TCP and the already discussed CoAP/UDP, support request response communications by defining means to link a given outgoing request to its reply. MQTT does not provide dedicated means to do such a linking. Thus we specify topics for both the request and the response, but it is responsibility of the programmer to ensure that corresponding topics are used in the client and in the server. A possibility for the programmer is to send the topic for the response inside the payload of the request message. We identify the topic for the reply with the `.aliasResponse` Protocol parameter. Like for `.alias` parameters, the template of the `.aliasResponse` parameter is instantiated using the content of the message sent in the behavior. For example, in Listing 6, we use `.id` in line 20 to obtain `"42/getTemperature"` and `"42/getTempReply"`, respectively the publication and reply topics.

We can now describe the pattern of interactions that we use to implement the outgoing `RequestResponse` communication at line 20 in Listing 6. As a reference, the pattern of interactions is depicted in the left part of Fig. 3. We will describe the right part later on, after having introduced inbound request-response communications.

First, the controller subscribes to the reply topic `"42/getTempReply"` at the Broker. Then, the controller sends to the Broker the request message on topic `"42/getTemperature"`. The execution of the `RequestResponse` terminates when the Broker forwards the reply received on topic `"42/getTempReply"` to the controller.

Differently from inbound `OneWay` communications, here we do not subscribe to the reply topic when the program is launched. Indeed, it would be useless since no relevant message can arrive on this topic before the controller sends its message to the Broker, and anticipating the subscription would complicate the usage of runtime information in templates.

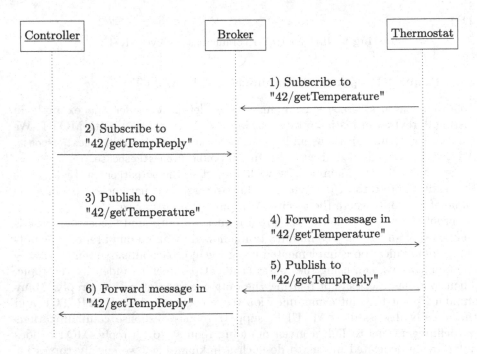

Fig. 3. Interaction in the temperature automation example in MQTT.

To exemplify inbound `RequestResponse` communications, we assume that the thermostat in our example is programmed in JIoT. We report its code in Listing 7.

```
1  interface ThermostatInterface {
2    RequestResponse: getTmp( TmpType )( TmpType )
3  }
4
5  inputPort Thermostat {
6    Location: "socket://localhost:9000"
7    Protocol: mqtt {
8      .broker = "socket://localhost:1883";
9      .osc.getTmp << {
10       .format = "raw",
11       .alias = "42/getTemperature",
12       .aliasResponse = "42/getTempReply"
13     }
14   }
15   Interfaces: ThermostatInterface
16 }
17
18 main {
19   //              ↓ receive the temperature and store it under the root of temp
20   getTmp( temp )( temp ){
21   //              ↑ update the content of temp and send it back as response
22   }
23 }
```

Listing 7. JIoT thermostat communicating over MQTT.

At line 11 in Listing 7, the `.alias` parameter `"42/getTemperature"` must be defined statically, as required for `inputPorts`. When the thermostat program is launched, it subscribes to topic `"42/getTemperature"`. When a message on this topic arrives, the payload (empty in this case) is passed to the behavior. The body of the `RequestResponse` (line 20) is executed to compute the return value. Finally, the return value is published on the reply topic `"42/getTempReply"`, as specified by osc parameter `.aliasResponse`. While in this example the parameter `.aliasResponse` is statically defined, our implementation supports the definition of dynamic `.aliasResponses` as in `outputPorts` (e.g., as seen in Listing 6).

We now summarize the exchange between the controller and the thermostat (left part of Fig. 3):

1. when the thermostat is started, it subscribes to topic `"42/getTemperature"` at the Broker;
2. when the outgoing `RequestResponse` is executed, the controller subscribes to topic `"42/getTempReply"` at the Broker;
3. the controller publishes the request message to topic `"42/getTemperature"`;
4. the Broker forwards the message in topic `"42/getTemperature"` to the thermostat;

5. the thermostat publishes the computed response at topic `"42/getTempReply"`;
6. the Broker forwards the message on topic `"42/getTempReply"` to the controller.

We remark that `RequestResponse` operations in Jolie are meant to be end-to-end communications. To ensure this in a publish/subscribe setting while using the approach above, one has to ensure that no other participant subscribes to the selected topics, which essentially act as namespaces.

7   Implementation

To illustrate the structure of our implementation, in Sect. 7.1 we discuss how media and protocols are separated from the Jolie interpreter and available as independent libraries. Then we describe the highlights of the implementation of UDP and CoAP in Sect. 7.2 and of MQTT in Sect. 7.3.

7.1   Programming a Jolie Extension

In Jolie the implementations of the supported application and transport protocols are independent. This enables the composition of any transport protocol with any application protocol. Concretely, the Jolie language is written in Java and provides proper abstract classes that represent application and transport protocols. Each protocol is obtained as an implementation of the corresponding abstract classes. Each implementation is a separated library which is loaded only if the protocol is used. This expedites the integration of new protocols in the language.

To better illustrate this structure, we report in Fig. 4 a conceptual representation of the call flow that originates from the execution logic of the language and interacts with the external libraries present in a given installation. The flow starts from the Execution Engine, which interprets Jolie commands, and which is the originator of the communication flows. This is represented by arrow ⓪ from the Execution Engine. From there, the call reaches the Communication Core, which implements the generic logic of channel creation, in turn relying on the pairing of a medium and a protocol. In the interpreter, this division is generalized with abstract factories for media and protocols. At runtime, the Communication Core proceeds (arrows ①) to load the medium factory requested in the call from the Execution Engine — in the figure we assume this is Socket — and, from that, it obtains an implementation of the actual logic of TCP/IP channels, split between a channel class, to handle outbound communications, and a listener class, for inbound communications. Finally, the Communication Core associates (arrows ②) a protocol to the obtained medium. The flow is similar to that of media: the Communication Core loads the protocol factory requested in the call from the Execution Engine — in the figure we assume this is HTTP — and, from that, it obtains an object that implements the logic of the HTTP protocol.

7.2 Implementation of CoAP/UDP in Jolie

Since by specification the CoAP protocol relies on the UDP medium protocol, in order to integrate CoAP in Jolie we also had to integrate the UDP medium. As described in Sect. 7.1, this entailed the creation of two new libraries for the Jolie interpreter: a medium library for UDP and a protocol library for CoAP.

We remark that, since UDP and CoAP are independent libraries, our implementation of UDP can also be used to support other protocols relying on UDP, such as MQTT-SN [26]. The implementation of UDP consists in a listener and a channel class, both based on the Netty framework [27]. Since the structure expected by Jolie and the one provided by Netty are similar, the integration of UDP is smooth. An interesting point is that exceptions raised by Netty are captured and transformed into Jolie exceptions. These exceptions are notified to the application protocol, which can either manage them or raise them at the level of the behavior of the Jolie program.

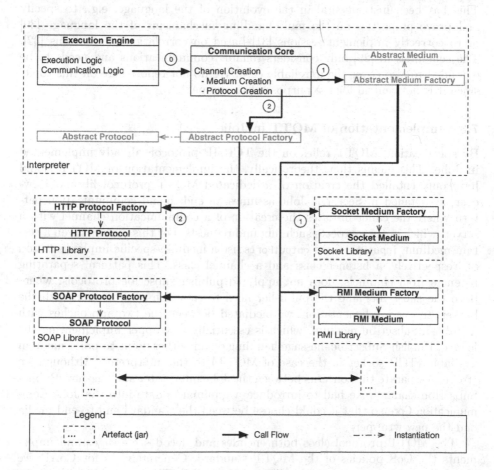

Fig. 4. Conceptual representation of the call flow among the Jolie interpreter and its communication libraries.

The implementation of the CoAP library consists in a class taking care of encoding and decoding the message abstraction of Jolie, namely the Communication Message, into a CoAP formatted one. A second class, handling the encoding and decoding of a CoAP message into a buffer of bytes, is based on the work done in nCoAP [28], an open source project providing a CoAP implementation for Java, based itself on Netty.

CoAP supports request-response communications and, in particular, CoAP messages include fields (i) to specify at which address the reply is expected and (ii) to match a reply with a previous request. Hence, the implementation of RequestResponse communications in CoAP is sound also with a transport protocol which is not connection-oriented, such as UDP. This would be a problem for protocols that do not provide such a facility, such as HTTP, which is indeed not commonly used over UDP.

Notably, Jolie comes with a formal semantics (in terms of a process calculus) [29], which enables to rigorously reason on the behavior of Jolie programs. This has been instrumental in the evolution of the language, e.g., to specify and prove properties on the fault handling mechanisms of the language [30] or to correctly implement sessions [31] based on correlation mechanisms [32]. The semantics in [29] only considers reliable communications and needs to be extended to also cover the unreliable case. We do not report here on this topic, since it is not central for the purpose of this paper.

7.3   Implementation of MQTT in Jolie

By specification, MQTT relies on the TCP/IP protocol, already implemented in Jolie. This means that, theoretically, the implementation of MQTT would have only entailed the creation of a dedicated MQTT protocol library. However, as detailed in Sect. 7.1, Jolie assumes an end-to-end communication pattern where the caller initiates the creation of a communication channel with a server, which in turn expects such inbound requests. For this reason, given a certain medium, inputPorts and outputPorts use a medium-specific implementation of, respectively, a listener class and a channel class. This pattern, separating listeners from channels, does not apply to publish/subscribe protocols, where both the subscriber and the publisher need to establish a connection with the broker. In our implementation, we mediated between the two approaches with a Publish-Subscribe medium, which is essentially a wrapper implementing the logic of Publish-Subscribe message handling on any other point-to-point medium available (TCP socket in the case of MQTT) to the interpreter. Although we strove to separate the concerns between the Jolie interpreter and this new Public-Subscribe channel, we had to introduce a minimal update into the Jolie Communication Core so that it could choose between the standard end-to-end media and the new wrapper.

The MQTT protocol class both encodes and decodes messages and implements the QoS policies of the MQTT standard. Concretely, as for CoAP, we based the implementation of MQTT on Netty [27]. The main difficulty in the implementation of the protocol is the definition of the message patterns needed

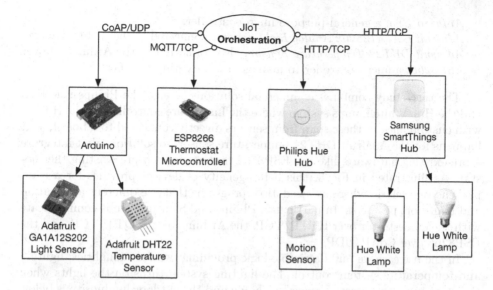

Fig. 5. Conceptual overview of the home automation case study.

to implement `OneWay` and `RequestResponse` communications, which have been described in Sect. 6. Beyond being invoked when operations are executed, the MQTT class is also invoked when the program is started, to perform port initialization. In particular, this is when subscriptions to topics identified in `inputPorts` are performed (along with the related connections to the brokers).

8   Case Study

In this section, we detail the programming of a home automation case study with JIoT. We remark that the techniques presented in this case study are not specific to home automation and can be used in any setting where heterogeneous IoT technology stacks need interact. The use case is peculiar as new edge devices can be included in the system at runtime. The source code of the system is released under the GPL v.3 license and available at [19]. We report in Fig. 5 a schematic overview of the case study, where Cloud nodes and mid-tier controllers (represented by the element labeled "JIoT Orchestration" in Fig. 5) are programmed in JIoT and orchestrate the behavior of a number of heterogeneous edge devices (whose low-level programming is omitted here):

– *Philips Hue Hub*: a hub to control the Philips Hue smart home devices;
– *Philips Hue White Lamps*: connected to the hub above;
– *Samsung SmartThings Hub*: a hub to control devices following the Smart-Things specification [33];
– *Samsung SmartThings Motion Sensor*: connected to the hub above and used as a presence sensor;

- *Arduino Uno*: a general-purpose microcontroller;
- *Adafruit GA1A12S202 Analog Light Sensor*: connected to the Arduino above;
- *Adafruit DHT22 Temperature Sensor*: also connected to the Arduino above;
- *ESP8266*: a microcontroller to manage a pre-existing thermostat.

The case study combines commercial solutions — e.g., the Philips Hue Hub and the Hue White Lamps system where the Lamps are controlled by the Hub — with custom ones — these span from sensors directly connected to a board, as it happens for the Adafruit DHT22 temperature sensor, to solutions that integrate a pre-existing hardware, like the ESP8266 that manages a pre-existing thermostat. As illustrated in Fig. 5, this heterogeneity of devices provides for a comprehensive scenario where we need JIoT programs that use different application and transport protocols. In particular, Philips and Samsung Hubs communicate with the orchestrator over HTTP/TCP, the Arduino over MQTT/TCP, and the ESP8266 over CoAP/UDP.

In the use case we build a simple logic providing two functionalities: lighting and temperature system control. The lighting system turns on the lights when the motion sensor detects someone at home and the outdoor luminosity is below some threshold. The temperature control checks the temperature and turns on the heating system when the temperature is below some threshold. The threshold has different values depending on whether someone is at home or not.

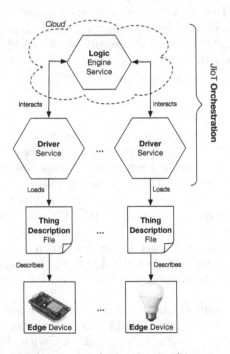

Fig. 6. Scheme of orchestration in the case study.

8.1   Structure of the Orchestration

We now describe the structure of the orchestration in the case study, which is illustrated in Fig. 6. The orchestration is composed of multiple JIoT programs. From top to bottom of Fig. 6, the LogicEngine contains the general logic of system control (i.e., the one that collects the data from sensors and coordinates the execution of the actuators in the system). Since the LogicEngine interacts with a multitude of mid-tier devices, its natural deployment is in the Cloud, where it is possible to scale it according to the number of managed devices and the load of computation. At the mid-tier level we have JIoT Drivers. Each Driver interacts with a specific edge device and it is deployed in a mid-tier machine in the proximity of the controlled edge device.

8.2   Thing Descriptors

In the case study, the Drivers are statically configured to manage a single fixed device using a JSON-LD 1.1 (that stands for JSON Linkage Data) configuration file [34]. The choice of JSON-LD is not mandatory, but it has the benefit of following the standard W3C Web of Things [22] definition of Thing Description (TD). This makes our Drivers already compliant with other WoT frameworks, simplifying future integrations with other WoT systems.

While discussing the full structure of TD is out of the scope of this paper, we present in Figs. 7 and 8 examples of TDs used in our case study. In Fig. 7 we report the TD for the DHT22 temperature sensor. For each device the JSON-LD file specifies whether it is a sensor or an actuator (key "type") and provides a textual description (key "description") and its name (key "name"). Each TD provides a list of properties (key "properties") that can be read. Each property is described by the property identifier, "temperature" in our example. The property identifier has various sub-elements describing it. In our example we use just key "label" to describe the unit of measure.

```
1  {
2    "type": "sensor",
3    "description": "Thing uses JSON-LD 1.1 serialization",
4    "name": "Adafruit DHT22 Temperature Sensor",
5    "properties": [
6      {
7        "temperature": {
8          "label": "Celsius"
9        }
10     }
11   ]
12 }
```

Fig. 7. Adafruit DHT22 Thing Descriptor.

```
1   {
2     "type": "actuator",
3     "description": "Thing uses JSON-LD 1.1 serialization",
4     "name": "Philips Hue White Lamp",
5     "actions": {
6       "toggleLight": {
7         "description": "Turn on or off the lamp."
8       }
9     }
10  }
```

Fig. 8. Philips Hue White Lamp Thing Descriptor.

JSON-LD configuration files for MQTT and HTTP devices are similar. Also configuration files for sensors and actuators are similar. As an example, we report in Fig. 8 the configuration file for Philips Hue White Lamps. There the main differences with respect to the previous TD (Fig. 7) are:

- the "type" is now "actuator";
- the key "actions" replaces the key "properties";
- the key "description" is used also to describe the single action.

In principle a TD can describe multiple properties belonging to a group of one or more edge devices controlled by the same Driver. For simplicity, here we have one TD for each edge device and, correspondingly, one Driver that controls one edge device. We also assume that each sensor provides one property.

8.3   System Deployment

Deployment-wise, JIoT provides a vast choice regarding what technology stack to use between the LogicEngine and the Drivers. Moreover, since both programs are developed in JIoT, it is easy to change their deployment, switching to the technology stack that best suites a given scenario (e.g., HTTP, to exploit caching, or binary formats like SODEP [17], to limit bandwidth usage). Here, we choose to use the HTTP/TCP stack to make our system compatible with the majority of existing third-party solutions [9]. However, different technology stacks fit different purposes. The benefit of JIoT is that programmers can re-use the same software components adapting their deployment to the desired communication stacks. For example, if our goal was to be natively compatible with other JavaScript IoT frameworks, we could have used the JSON-RPC binary protocol; if we wanted to deploy our system as part of a Service-Oriented Architecture [14], we could have used the SOAP protocol.

While JIoT-to-JIoT deployment is flexible, the deployment towards edge devices is defined by the technology supported by the edge device. Concretely, in our case study each Driver communicates with its edge device using (one of) the protocol(s) supported by the latter.

8.4   Components Behavior

When started, a Driver loads the TD configuration file of its edge device. Then it registers itself to the LogicEngine. In the registration, it sends the information retrieved from the TD, enriched with two additional pieces of information: the address where the edge device can be contacted — i.e., the Driver location — and the identifier of the user to which the edge device belongs. Once registered, the Driver acts as a forwarder between the LogicEngine and the edge device.

The LogicEngine runs on the Cloud and manages a number of sensors and actuators. More precisely the LogicEngine has one running session for each user (distinguished according to the user identifier), managing all her sensors and actuators. Each session is associated with an array of devices that can be scanned to find the location of devices with specific properties and interact with them; e.g., at lines 10–26 of Listing 8 the procedure getTemperature of the LogicEngine, computing the average temperature recorded by the sensors of one user.

```
1  interface driverInterface {
2    RequestResponse: engineRequest( request )( response )
3  }
4
5  outputPort Driver {
6    Protocol: http
7    Interfaces: driverInterface
8  }
9
10 define getTemperature {
11   sum = 0 ;
12   n = 0 ;
13   for ( device in devices ) {
14    if( device.type == "sensor" &&
15      is_defined( device.properties.temperature ) ) {
16     Driver.location = device.driverLocation ;
17     request.operationName = "getTemperature" ;
18     engineRequest@Driver( request )( response ) ;
19     sum = sum + response.deviceResponse ;
20     n++
21    }
22   } ;
23   if( n!=0 ) {
24    temperature = sum / n
25   }
26 }
```

Listing 8. LogicEngine Driver **outputPort** and getTemperature procedure.

Briefly, procedure getTemperature:

- scans the devices structure (line 13) containing all registered Drivers;
- selects those whose type is "sensor" and have a property (under the substructure properties) named temperature. Note how Jolie tree-shaped variables ease the exploration of structured data; in this case the one sent by the Drivers at registration time (and read from their associated JSON-LD file);
- it dynamically sets (line 16) the location of outputPort Driver (lines 5–8) to contact the selected Driver;
- it sets the request operation to getTemperature (line 17);
- it retrieves the temperature sensed by the edge device controlled by the selected Driver, invoking it through operation engineRequest;
- it aggregates the sensed temperature in variable sum and keeps track of the number of requests in variable n (lines 19–20);
- it computes the mean temperature (lines 23–25).

The procedures that calculate the mean of the sensed external luminosity and the one to check the presence of people at home are similar to the one in Listing 8, except that the searched properties are light in the first case, and motion in the second.

We report in Listing 9 one of the procedures managing the actuators, specifically the one used to set the temperature. The main difference with respect to the logic in Listing 8 is that procedure setTemperature:

- selects the devices whose type is "actuator" (line 3);
- sets the request operation to "setTemperature" and passes the value in variable comfortTemperature as parameter of the request (lines 6–7).

Note that the operation called on the Driver is engineRequest both in Listing 8 and Listing 9. This support the extension of the LogicEngine with new

```
1 define setTemperature {
2  for ( device in devices ) {
3   if( device.type == "actuator" &&
4     is_defined( device.properties.temperature ) ) {
5    Driver.location = device.location ;
6    request.operationName = "setTemperature" ;
7    request.deviceRequest = comfortTemperature ;
8    engineRequest@Driver( request )( response )
9   }
10  }
11 }
```

Listing 9. LogicEngine setTemperature procedure.

procedure definitions that implement a given goal without requiring to change the interface between the LogicEngine and the Drivers. In turn, a request with the same operationName (e.g., "setTemperature") triggers different behaviors in different Drivers, as each implements the specific logic of interaction with its associated edge device.

8.5   Cloud Deployment

We conclude this section by describing the Cloud deployment of the LogicEngine, which is containerized using Docker [35]. The container is deployed automatically into an Amazon Web Service cluster via the Docker Swarm manager [36]. The LogicEngine is deployed in the worker cluster, allowing the manager to balance the load of requests. We report in Listing 10 the content of the Dockerfile used to deploy the LogicEngine.

At line 1 we declare the starting image for the container, which is the lightweight Linux Alpine distribution with OpenJDK 8 pre-installed. At lines 3–4 we install the JIoT fork interpreter and we set the environmental variable JOLIE_HOME to point to the location of the installed interpreter. At lines 6–7 we add the source code of the LogicEngine in the home directory of the image. Finally, at line 8 we start the execution of the LogicEngine.

9   Related Work

In the literature there are many proposals for platforms, middlewares, smart gateways, and general systems, all aimed at solving the interoperability problem arising from the current "babel" of IoT technologies (protocols, formats, and languages). Without any claim of being complete, here we mention a few notable examples which are somehow related to the topic of the current paper.

Recently the W3C started the Web of Things (WoT) Working Group [22]. The aim of WoT is to define a standard stack of layered technologies, as well as software architectural styles and programming patterns, to uniform and simplify the creation of IoT applications. In this context, the W3C is working on

```
1 FROM openjdk:alpine
2
3 RUN java -jar jiot.jar -jh /usr/lib/jolie/ -jl /usr/bin/
4 ENV JOLIE_HOME /usr/local/lib/jolie
5
6 ADD logic_engine.ol /home/.
7 WORKDIR /home
8 RUN jolie logic_engine.ol
```

Listing 10. The Dockerfile used to deploy the LogicEngine.

a WoT Architecture [37]. The main concept of the architecture is the notion of "servient", a virtual entity that represents a physical IoT device. Servients provide technology-independent, standard APIs that developers can use to transparently operate in heterogeneous environments. Remarkably, both the WoT proposal and ours concern high-level abstractions for low-level access to devices provided via, e.g., HTTP, CoAP, and MQTT. However, while we propose a dedicated language, they provide API specifications. More in general, there are many proposals for the integration of WoT and IoT. For example [38] and [39] define general platforms covering different layers of IoT, including an accessibility layer which integrates concepts like smart gateways and proxies to facilitate the connection of (smart) Things into the Internet infrastructure, using architectural principles based on REST. Smart gateways and proxies are used in several industrial proposals to facilitate the development of applications. Common denominator of some of these proposals, e.g., [33,40,41], is the abstraction of low-level functionalities provided by embedded devices (e.g., connectivity and communication over low-level protocols like ZigBee, Z-Wave, Wi/IP/UPnP, etc.). Smart gateways are used also to translate (or integrate) CoAP into HTTP [42–44] and to integrate both CoAP and MQTT by means of specific middlewares [45]. Eclipse IoT [46] is an IoT integration framework proposed by the Eclipse IoT Working Group. Aim of Eclipse IoT is to build an open IoT stack for Java, including the support for device-to-device and device-to-server protocols, as well as the provision of protocols, frameworks, and services for device management. There exist several European projects, notably INTER-IoT [47] and symbIoTe [12], that address the issue of interoperability in IoT and have produced several concrete proposals. Finally, a work close to ours is [48], where a middleware converts IoT heterogeneous networks into a single homogeneous network.

Although related to our aim in this paper, the cited proposals tackle the problem of IoT integration from a framework perspective: they provide chains of tools, each addressing a specific level of the integration stack. Differently, we extend a language specifically tailored for system integration and advanced flow manipulation, Jolie, to support integration of IoT devices. This offers a single linguistic domain to seamlessly integrate disparate low-level IoT devices and intermediate nodes (collectors, aggregators, gateways). Moreover, Jolie is already successfully used for building Cloud-based, microservice solutions [49,50]. This makes the language useful also for assembling advanced architectures for IoT, e.g., to handle real-time streaming and processing of data from many devices. The benefit, here, is that, while solutions based on frameworks require dedicated proficiencies on each of the included tools, Jolie programmers can directly work at any level of the IoT stack, without the need to acquire specific knowledge on the tools in a given framework.

To conclude our revision of related work, we narrow our focus on language-based integration solutions for IoT. The work mostly related to ours is SensorML [51]. SensorML, abbreviation of Sensor Model Language, is a modeling language for the description of sensors and, more in general, of measurement processes. Some features modeled by the language are: discovery and geolocalization of sensors, processing of sensor observations, and functionalities to program sensors and to subscribe to sensor events. While some traits of SensorML are common to our proposal, the scopes of the two languages sensibly differ. Indeed, while Jolie is a high-level language for programming generic architectures (spanning from Cloud-based microservices to low-level IoT integrators), SensorML just models IoT devices, their discovery, and the processing of sensor observations.

10 Discussion and Conclusion

In this paper, we proposed a language-based approach for the integration of disparate IoT platforms. We built our treatment on the Jolie programming language. This first result is an initial step towards a more comprehensive solution for IoT ecosystem integration and management. Concretely, we included in Jolie the support for two of the most widely used IoT protocols. The inclusion enables Jolie programmers to interact with the majority of present IoT devices. Summarizing our results: (*i*) we included in Jolie the CoAP application protocol, also extending the Jolie language to support the UDP transport protocol, (*ii*) we added the support for the MQTT protocol and, in doing so, (*iii*) we tackled the challenging problem of mapping the renowned pattern of request-responses (typical of HTTP and other widely used protocols) into the publish/subscribe message pattern of MQTT. The mapping abstracts from peculiarities of MQTT and is applicable to any publish/subscribe protocol.

Regarding future work, we are currently investigating the integration in Jolie of more IoT protocols [3], in order to extend the usability of the language in the IoT setting.

It would also be interesting to extend not only the Jolie interpreter, as we have done, but also the formal model behind it [29,31,52]. To this end, we can take ideas from the formal model of IoT systems presented in [53].

Another interesting direction for future developments is studying how Jolie can support the testing of IoT technologies, e.g., to test how different protocol stacks perform over a given IoT topology. Thanks to the simplicity of changing the combination of the used protocols (application and transport), experimenters can quickly test many configurations, also enjoying a more reliable platform to compare them. Indeed, usually even changing one of the protocols in the configured stack would require an almost complete rewrite of the logic of network components. Contrarily, in Jolie, this change just requires an update of the deployment part of programs, leaving the logic unaffected. Moreover, such an update could even be done programmatically, making the practice of repeated experimenting on IoT networks easier and more standardized.

Finally, as future work, we also consider the possibility of developing a lightweight version of the language, to be used on low-power IoT devices. Indeed, in this paper, we assumed that these devices are programmed with low-level languages, since they can support only a very constrained execution environment. Clearly, letting programmers develop all the components of an IoT network in the same language would not only ease its implementation but also testability, deployment, and maintenance. However, achieving such a result would require a very challenging engineering endeavor.

Acknowledgments. We thank Marco Di Felice, Luca Bedogni, and Federico Montori for useful suggestions and comments.

References

1. Gubbi, J., Buyya, R., Marusic, S., Palaniswami, M.: Internet of Things (IoT): a vision, architectural elements, and future directions. Future Gener. Comput. Syst. **29**(7), 1645–1660 (2013)
2. Atzori, L., Iera, A., Morabito, G.: The Internet of Things: a survey. Comput. Netw. **54**(15), 2787–2805 (2010)
3. Al-Fuqaha, A., Guizani, M., Mohammadi, M., Aledhari, M., Ayyash, M.: Internet of Things: a survey on enabling technologies, protocols, and applications. IEEE Commun. Surv. Tutorials **17**(4), 2347–2376 (2015)
4. Bormann, C.: CoAP website. http://coap.technology/ (2016)
5. Shelby, Z., Hartke, K., Bormann, C.: The constrained application protocol (CoAP), RFC 7252, IETF (2014)
6. MQTT community: MQTT website. http://mqtt.org (2014)
7. Banks, A., Gupta, R.: MQTT Version 3.1.1, Oasis standard, Oasis (2014). http://docs.oasis-open.org/mqtt/mqtt/v3.1.1/os/
8. Milenkovic, M.: A case for interoperable IoT sensor data and meta-data formats: the Internet of Things (Ubiquity symposium). Ubiquity **2015**, 2:1–2:7 (2015)
9. Richardson, L., Ruby, S.: RESTful Web Services. O'Reilly Media Inc, Newton (2008)
10. Garg, N.: Apache Kafka. Packt Publishing Ltd, Birmingham (2013)
11. Soursos, S., Žarko, I.P., Zwickl, P., Gojmerac, I., Bianchi, G., Carrozzo, G.: Towards the cross-domain interoperability of IoT platforms. In: EuCNC, pp. 398–402. IEEE (2016)
12. Gojmerac, I., Reichl, P., Podnar Žarko, I., Soursos, S.: Bridging IoT islands: the symbIoTe project. Elektrotechnik und Informationstechnik **133**(7), 315–318 (2016)
13. The bIoTope project. http://www.biotope-project.eu/ (2017)
14. Erl, T.: Soa: Principles of Service Design. Prentice Hall Press, New Jersey (2007)
15. Montesi, F., Guidi, C., Lucchi, R., Zavattaro, G.: JOLIE: a java orchestration language interpreter engine. ENTCS **181**, 19–33 (2007)
16. Montesi, F., Guidi, C., Zavattaro, G.: Composing services with JOLIE. In: ECOWS, pp. 13–22. IEEE (2007)
17. Montesi, F., Guidi, C., Zavattaro, G.: Service-oriented programming with Jolie. Web Services Foundations. Springer, New York (2014). https://doi.org/10.1007/978-1-4614-7518-7_4
18. Jolie website. http://jolie-lang.org (2017)
19. Gabbrielli, M., Giallorenzo, S., Lanese, I., Zingaro, S.P.: Jolie for IoT website. http://www.cs.unibo.it/projects/jolie/jiot.html (2017)
20. W3C: Transport message exchange pattern: single-request-response. https://www.w3.org/2000/xp/Group/1/10/11/2001-10-11-SRR-Transport_MEP (2001)
21. Gabbrielli, M., Giallorenzo, S., Lanese, I., Zingaro, S.P.: A language-based approach for interoperability of IoT platforms. In: HICSS, AIS Electronic Library (AISeL) (2018)

22. Web of Things. https://www.w3.org/WoT/ (2017)
23. Fielding, R.T.: Architectural styles and the design of network-based software architectures. PhD thesis, University of California, Irvine (2000)
24. Postel, J.: User datagram protocol. RFC 768, IETF (1980)
25. Montesi, F.: Process-aware web programming with Jolie. SCP **130**, 69–96 (2016)
26. Hunkeler, U., Truong, H.L., Stanford-Clark, A.: MQTT-S–a publish/subscribe protocol for wireless sensor networks. In: COMSWARE, pp. 791–798. IEEE (2008)
27. Maurer, N., Wolfthal, M.: Netty in Action. Manning Publications, New York (2016)
28. Kleine, O.: nCoAP. https://github.com/okleine/nCoAP
29. Guidi, C., Lucchi, R., Gorrieri, R., Busi, N., Zavattaro, G.: SOCK: a calculus for service oriented computing. In: Dan, A., Lamersdorf, W. (eds.) ICSOC 2006. LNCS, vol. 4294, pp. 327–338. Springer, Heidelberg (2006). https://doi.org/10.1007/11948148_27
30. Guidi, C., Lanese, I., Montesi, F., Zavattaro, G.: Dynamic error handling in service oriented applications. Fundam. Inform. **95**(1), 73–102 (2009)
31. Montesi, F., Carbone, M.: Programming services with correlation sets. In: Kappel, G., Maamar, Z., Motahari-Nezhad, H.R. (eds.) ICSOC 2011. LNCS, vol. 7084, pp. 125–141. Springer, Heidelberg (2011). https://doi.org/10.1007/978-3-642-25535-9_9
32. OASIS: Web Services Business Process Execution Language. http://docs.oasis-open.org/wsbpel/2.0/wsbpel-v2.0.html
33. SmartThings. http://www.smartthings.com/ (2016)
34. Sporny, M., Longley, D., Kellogg, G., Lanthaler, M., Lindström, N.: JSON-LD 1.1. https://json-ld.org/spec/latest/json-ld/
35. Merkel, D.: Docker: lightweight linux containers for consistent development and deployment. Linux J., vol. 2014, Mar 2014
36. Soppelsa, F., Kaewkasi, C.: Native Docker Clustering with Swarm. Packt Publishing, Birmingham (2017)
37. Web of Things architecture. https://w3c.github.io/wot/architecture/wot-architecture.html (2017)
38. Dominique, G.: A web of things application architecture-integrating the real-world into the web. Zurich Diss. ETH **19891**, 10–12 (2011)
39. Corredor, I., Metola, E., Bernardos, A.M., Tarrío, P., Casar, J.R.: A lightweight web of things open platform to facilitate context data management and personalized healthcare services creation. IJERPH **11**(5), 4676–4713 (2014)
40. Meshlium. http://www.libelium.com/products/meshlium/ (2016)
41. Thinking things. http://www.thinkingthings.telefonica.com/ (2016)
42. Sulaeman, A.B., Ekadiyanto, F.A., Sari, R.F.: Performance evaluation of HTTP-CoAP proxy for wireless sensor and actuator networks. In: APWiMob, pp. 68–73, IEEE (2016)
43. Ludovici, A., Calveras, A.: A proxy design to leverage the interconnection of CoAP wireless sensor networks with web applications. Sensors **15**(1), 1217–1244 (2015)
44. Mingozzi, E., Tanganelli, G., Vallati, C.: CoAP proxy virtualization for the Web of Things. In: CloudCom, pp. 577–582. IEEE Computer Society (2014)
45. Thangavel, D., Ma, X., Valera, A., Tan, H.X., Tan, C.K.Y.: Performance evaluation of MQTT and CoAP via a common middleware. In: ISSNIP, pp. 1–6. IEEE (2014)
46. The Eclipse for IoT Project. https://iot.eclipse.org/ (2017)
47. Ganzha, M., Paprzycki, M., Pawlowski, W., Szmeja, P., Wasielewska, K.: Semantic technologies for the IoT - an inter-IoT perspective. In: IoTDI, pp. 271–276. IEEE (2016)

48. Zhiliang, W., Yi, Y., Lu, W., Wei, W.: A SOA based IoT communication middle-ware. In: MEC, pp. 2555–2558. IEEE (2011)
49. Gabbrielli, M., Giallorenzo, S., Guidi, C., Mauro, J., Montesi, F.: Self-reconfiguring microservices. In: Ábrahám, E., Bonsangue, M., Johnsen, E.B. (eds.) Theory and Practice of Formal Methods. LNCS, vol. 9660, pp. 194–210. Springer, Cham (2016). https://doi.org/10.1007/978-3-319-30734-3_14
50. Callegati, F., Giallorenzo, S., Melis, A., Prandini, M.: Insider threats in emerging mobility-as-a-service scenarios. In: HICSS, AIS Electronic Library (AISeL) (2017)
51. The sensorML project. http://www.opengeospatial.org (2017)
52. Giallorenzo, S., Montesi, F., Gabbrielli, M.: Applied choreographies. In: Baier, C., Caires, L. (eds.) FORTE 2018. LNCS, vol. 10854, pp. 21–40. Springer, Cham (2018). https://doi.org/10.1007/978-3-319-92612-4_2
53. Lanese, I., Bedogni, L., Di Felice, M.: Internet of Things: a process calculus app-roach. In: SAC, pp. 1339–1346. ACM (2013)

Energy-Efficient Scheduling of Tasks with Conditional Precedence Constraints on MPSoCs

Umair Ullah Tariq[1], Hui Wu[1(✉)], and Suhaimi Abd Ishak[2]

[1] The University of New South Wales, Sydney, Australia
{u.tariq,huiw}@unsw.edu.au
[2] Universiti Tun Hussein Onn, Parit Raja, Malaysia
suhaimiabd@uthm.edu.my

Abstract. In this article, we investigate the problem of energy-efficient scheduling of tasks with conditional precedence constraints on heterogeneous NoC-based MPSoC. We propose a novel offline approach that performs task mapping, scheduling and voltage scaling in an integrated manner. Our approach consists of a scheduling algorithm that constructs a single unified schedule by prioritizing tasks with tight latest finish time bounds. It uses an NLP-based DVFS algorithm to assign continuous frequencies and voltages to tasks and communications, and transforms the assigned frequencies and voltages to tasks and communications to valid discrete frequency and voltage levels using either an ILP or a heuristic-based algorithm. Compared to the state-of-the-art approach designed for the task model with unconditional precedence constraints, our approach using ILP-based algorithm achieves improvements in the range of 9% to 61% and an average improvement of 31%, and our approach using a heuristic-based algorithm achieves improvements in the range of 2% to 46% and an average improvement of 20% in terms of energy reduction. In terms of running time, our approach is approximately 3 times faster than the state-of-the-art approach.

Keywords: Conditional Task Graph (CTG) ·
Dynamic voltage and frequency scaling ·
Offline task mapping and scheduling

1 Introduction

During the past few decades, we have witnessed immense growth in the applications of embedded systems. In addition to conventional performance metrics such as execution speed, energy consumption is a critical metric for gauging the quality of an embedded system. Modern embedded systems such as driverless cars and robots require powerful energy-efficient hardware due to their complex functions. Multi-Processor System on Chip (MPSoC) is an ideal architecture for

© Springer Nature Switzerland AG 2019
T. A. Majchrzak et al. (Eds.): *Towards Integrated Web, Mobile,*
and IoT Technology, LNBIP 347, pp. 115–145, 2019.
https://doi.org/10.1007/978-3-030-28430-5_6

these systems due to its high performance and low power dissipation. MPSoC has knobs that provide a mechanism to trade-off speed of execution against energy efficiency [39]. Adjusting the settings of these energy-management knobs so that the impact on execution speed is minimized and the energy reduction is maximized, involves decisions such as where, when and at what speed to execute each task of an application. These decisions are made by the task scheduling algorithm. Therefore, energy-aware task scheduling is an important research topic.

Typically an embedded application consists of a set of tasks. Each task represents a piece of code in the application. Tasks may be subject to precedence constraints. A precedence constraint (A, B) between tasks A and B specifies that task B can start only after A has finished. Classic Directed Acyclic Graph (DAG) task model captures such relationships between tasks. In applications some precedence constraints may be conditional. A conditional precedence constraint (A, B) specifies that task B can start only after A has completed only if a certain condition has been met. Applications with conditional precedence constraints are modeled by Conditional Task Graphs (CTG). A real example of an application with conditional precedence constraints is an MPEG decoder. An MPEG decoding process varies according to the frame of the encoded video stream. Each frame is composed of 16×16 pixel macro-blocks. Macro-blocks are classified as I, P and B blocks. Each macro block has a different decoding procedure. For example I block performs Inverse Discrete Cosine Transform (IDCT) but B-block may skip IDCT. Traditional DAG-based task models cannot model such an application.

Scheduling applications on MPSoCs involve task mapping that allocates tasks to MPSoC processors, and task ordering that arranges tasks in time. In heterogeneous MPSoCs processors have not only different power-performance characteristics but may also have different voltage scaling capabilities. In such systems task mapping greatly influences the capability of DVFS algorithms to reduce energy consumption. DVFS algorithms are also influenced by the order in which tasks are executed as task executions are bounded by deadlines. If tasks are not mapped on energy-efficient processors and tasks with shorter deadlines block tasks with longer deadlines, DVFS can do very little to reduce energy consumption. Moreover, modern MPSoCs have a large number of processors and for the future kilo-processor MPSoCs [12] bus-based on-chip communication is no longer feasible due to its poor scalability. Network on Chip (NoC)-based on-chip interconnects are suitable for these MPSoCs as they provide a significant improvement in terms of flexibility, scalability and performance over other on-chip interconnect such as hierarchical (e.g., advanced micro-controller bus architecture and STBus) and traditional bus structures [20].

In this paper we investigate the problem of scheduling a set of tasks with conditional precedence constraints and individual deadlines on a heterogeneous NoC-based MPSoC such that the total expected processor and communication energy is minimized. The processors and NoC links are voltage scalable and can operate at a set of discrete voltage/frequency levels. We make the following major contributions:

1. We propose a novel offline task scheduling approach. Our approach consists of a task scheduling heuristic that constructs a single unified schedule for all the tasks and collectively assigns a frequency to each task and each communication assuming continuous frequencies, and an Integer Linear Programming (ILP)-based algorithm as well as a polynomial-time heuristic for assigning a discrete frequency to each task and each communication. To the best of our knowledge, our approach is the first one that investigates the problem of scheduling a set of tasks and communications with conditional precedence constraints on NoC-based MPSoCs such that the total expected energy consumption is minimized.

2. We have performed experiments on 20 benchmarks. Compared to the state-of- the-art approach proposed by Li and Wu [22] that does not consider conditional precedence constraints, in terms of energy reduction, our approach using the ILP-based algorithm achieves an average improvement of 31% and a maximum improvement of 61%, and our approach using the polynomial-time heuristic achieves an average improvement of 20% and a maximum improvement of 46%. Furthermore, both our approach using the ILP-based algorithm and our approach using the polynomial-time heuristic run approximately three times faster than the state-of-the-art approach.

2   Related Work

In NoC-based MPSoCs energy is not only consumed by the processors but also by the communication network. The energy (both static and dynamic) consumed by the processors and communication network is referred to as processing and communication energy, respectively. Authors in [27,33,40,41] schedule tasks with precedence constraints such that communication energy is reduced. These approaches achieve this goal through task duplication. The key idea of these approaches is to duplicate some tasks to reduce the communication energy as well as traffic congestion. Authors in [33] additionally take into account the contention among communications for communication links bandwidth. Although these approaches may optimize energy consumption but are limited in only reducing the communication energy, and therefore are only suitable for applications with intensive communication volumes.

Many energy-aware approaches have been proposed for scheduling tasks with precedence constraints with an objective to optimize the processing energy consumption.

Authors in [14] investigate the problem of scheduling a set of tasks with precedence constraints onto heterogeneous multi-processors system such that the dynamic energy consumption is minimized. In their approach task mapping and scheduling are integrated with dynamic voltage and frequency scaling to maximize energy efficiency. Their DVFS algorithm assigns task frequencies and voltages based on the critical path length of the tasks.

Authors in [28] propose DVFS-based offline and online algorithms for assigning frequencies to a set of tasks with precedence constraints on multi-processors systems. They assume that a static schedule is given, which assigns tasks to processors and specifies the order in which tasks are executed. They propose three off-line DVFS algorithms, Greedy Static Power Management (G-SPM), Simple Static Power Management (S-SPM) and Static Power Management with Parallelism (P-SPM). The G-SPM allocates the global slack (difference of schedule makespan and application deadline) to the first task on each processor, if the first task is a source node. The S-SPM allocates the global slack to tasks proportionally based on their worst-case execution times at maximum processor frequency. P-SPM first determines the degree of parallelism of the different time intervals of static schedule and allocates slack to each time interval based on the degree of parallelism of that interval (number of tasks scheduled in the interval). They also propose a greedy online DVFS algorithm that allocates the slack available at run-time to next ready task.

Authors in [5] propose offline voltage scaling algorithms to schedule tasks with precedence constraints on a heterogeneous multi-processor system such that the total energy consumption is minimized. They assume that a static schedule is given and propose only voltage scaling algorithms. For processors that can operate at any frequency and voltage within a continuous range, they propose a Non-Linear Programming (NLP)-based voltage scaling algorithm to assign each task a voltage and a frequency. For a processor that can operate at a fixed set of discrete voltages and frequencies, they propose an ILP-based voltage scaling algorithm. Their approaches optimize the total energy consumption by collectively scaling supply voltage and body-to-source voltage. Authors in [8] propose an energy-efficient approach to schedule tasks with precedence constraints on battery-powered heterogeneous multi-processor mobile embedded systems. They assume that task mapping and scheduling are given and propose an online algorithm that assigns frequencies and voltages to tasks. Their DVFS algorithm is based on the critical path length of the tasks and allocates a slack to each task proportional to the critical path length of the task.

Authors in [19] propose an online DVFS algorithm to schedule tasks with precedence constraints on a multi-processor system such that dynamic energy consumption is minimized. They assume that tasks have been mapped, scheduled and assigned frequencies and voltages offline, and propose online DVFS algorithms, including Greedy, K time lookahead and K descendant lookahead. The online DVFS algorithm is called at run-time whenever a task finishes execution. The greedy algorithm adjusts the frequency of only the direct successors of the finished task. K time lookahead algorithm adjusts the frequency of each task in a fixed time window k, and k descendant algorithm adjusts the frequencies of only k direct decedents of the finished task.

Authors in [2] investigate the problem of energy-aware scheduling of task with precedence and deadline constraint on heterogeneous multiprocessor systems. They assume that a static schedule is given and propose a voltage scaling algorithm called Quasi-Static Voltage Scaling (QSVS). The QSVS aims to minimize the overhead of online voltage scaling algorithm. They achieve this goal by distributing the voltage scaling process in offline and online phases. Their offline algorithm prepares the voltage settings and stores them in a look-up table. The online algorithm selects the voltage setting based on the workload at run-time.

Authors in [34] propose a heuristic called Enhanced Energy efficient Scheduling (EES) to schedule tasks with precedence and deadline constraints on a homogeneous multi-processor system such that energy consumption is minimized. The EES algorithm consists of three phases. The first phase is the task mapping phase in which tasks are mapped to processors using HEFT algorithm [36]. In the second phase, the slack between the makespan of the schedule and the deadline is distributed proportionally among the tasks on the critical path. In the final phase, the EES algorithm assigns the most energy-efficient voltages and frequencies to tasks recursively.

Authors in [23] schedule a set of tasks with precedence constraints on homogeneous multi-processor system such that timing constraints are satisfied and energy consumption is minimized. Their approach is based on level-by-level scheduling and has three main steps. In the first step, tasks are grouped into levels. All the tasks at the same level are independent. In the second step, each level is assigned a time-slot such that all tasks in the level execute in that time-slot. Finally, all the tasks in the same level are assigned the same frequency. Authors in [29] and [6] survey in detail approaches that optimize the processing energy consumption.

All the approaches discussed so far aim to minimize either the processing energy or the communication energy only. Next, we discuss some approaches that optimize both the processing energy and the communication energy. Authors in [21] propose a complete solution of task mapping, scheduling and voltage assignment for a task set with dependencies. They formulate the entire problem of task mapping, ordering and voltage assignment as a Mixed Integer Linear Programming (MILP) problem. Since MILP does not scale well, they also propose a divide-and-conquer-based polynomial-time heuristic algorithm.

Authors in [31] propose an approach that combines task mapping, scheduling, and DVFS to optimize the energy consumption. They consider a set of tasks with precedence constraints and heterogeneous multi-processor systems where each processor is DVFS-enabled and can operate at any frequency in a continuous range. Their approach consists of two nested Genetic Algorithms. The outer Genetic Algorithm performs task mapping and the inner Genetic Algorithm performs task ordering. Their DVFS algorithm assigns voltages to tasks such that timing constraints are satisfied and the energy consumption is minimized for the task mapping and task ordering generated by Genetic Algorithms. Authors in [16] propose an energy-aware task mapping, scheduling and voltage scaling algorithm for a task set with precedence constraints. They integrate task mapping, scheduling and voltage assignment in a single optimization loop. Their algorithm starts with an initial solution generated by a list scheduling algorithm assuming that all processors operate at maximum frequencies and iteratively tries to improve the initial solution. In each iteration, their algorithm tentatively re-maps each task to every processor at each voltage level of the processor and constructs a new schedule for the new mapping and voltage assignment, and finds a task with the highest priority (ratio of the difference in energy consumption and the difference in make-span) and re-maps it to the processor at a voltage level such that the energy consumption of the new solution is reduced compared to the current solution.

Authors in [15] investigate the problem of energy-efficient scheduling of tasks with precedence constraints on heterogeneous NoC-based MPSoC. They formulate the entire problem of task mapping, task and communication ordering and task voltage assignment as an MILP. They also propose a heuristic algorithm called randomized rounding. The randomized rounding heuristic algorithm first solves a Linear Programming (LP) problem by relaxing the integer constraints in MILP. It then repeatedly rounds the variables with non-integer values to integer values such that constraints are not violated.

Authors in [17] propose a simulated annealing-based, energy-aware task mapping algorithm on heterogeneous NoC-based MPSoCs. In their model processors are assumed to be voltage scalable and NoC links operate at a fixed frequency. They propose MILP-based algorithm that takes into account both communication and processing energy. They integrate a simulated the annealing algorithm with a timing adjustment heuristic. The timing adjustment heuristic explores the solution space near an acceptable mapping generated by the simulated annealing algorithm to find a mapping that reduces the energy consumption.

Authors in [15–17, 21, 31] aim to minimize both communication and processing energy. All these approaches assume that only processors are voltage scalable. Therefore, the DVFS approaches allocate the slack to tasks only and the communication energy is reduced only through task mapping. Authors in [4] and [3] have shown that if both processors and communication links are voltage scalable, more energy can be saved by sharing the available slack between communications and tasks.

Authors in [4] and [3] propose an NLP and an MILP-based DVFS algorithms for a tasks set with precedence constraints on heterogeneous MPSoC. Their proposed approach shares available slack between task and communication nodes such that total energy consumption is minimized. Authors in [32] consider a NoC based MPSoC model with voltage scalable links and assume that processors operate at fixed frequency and voltage levels. They propose energy efficient voltage scaling algorithm that aims to minimize the communication energy by statically assigning voltages and frequencies to links.

Authors in [22] propose task mapping, scheduling and DVFS algorithm for a task set with precedence constraints on homogeneous NoC based-MPSoC model with voltage scalable links and processors. They propose a two-step approach. In the first step, they propose a quadratic programming-based mapping algorithm that maps tasks to a processor such that total weighted communication distance is minimized. In the second step, they use GA to assign voltages and frequencies to tasks and communications.

Our approach differs from all the previous approaches in three major aspects. First, our approach considers conditional precedence constraints. Second, our approach handles NoC and takes link contentions into account. Third, our approach collectively optimizes the frequencies of processors and NoC links aiming at minimizing the total expected energy consumption of the MPSoC.

3   Models

The target application is modelled by a conditional task graph (CTG) [35]. A CTG is a weighted directed acyclic graph $G(V, E, A, X)$ defined as follows. $V = \{v_1, v_2, ..., v_n\}$ is a set of tasks. Each task has an execution time represented by the number of clock cycles on each processor and a deadline d_i. All the tasks are non-preemptible. $E \subset V \times V$ is a set of directed edges each denoting the dependency between the two tasks. A is a set of triplets $(e_i, c_i, p(c_i))$, where $e_i \in E$, and c_i and $p(c_i)$ represent the condition associated with e_i and its probability [24], respectively. X is a set of edge weights. An edge weight $X_s \in X$ of an edge $e_s = (v_i, v_j)$ represents the communication volume in bits from task v_i to task v_j. The execution probability of each node $v_i \in V$ is represented by $p(v_i)$ [24].

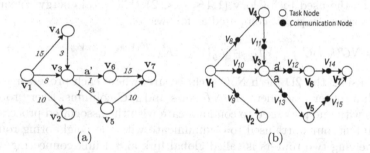

Fig. 1. (a) A CTG G (b) The extended CTG G_e

Each non-sink node of a conditional task graph is a FORK node. A FORK node with multiple outgoing edges is an OR-FORK node if all the outgoing edges are conditional edges with mutually exclusive conditions. The sum of the probabilities of all the conditional edges of an OR-FORK node is equal to 1. A node is an AND-FORK node if all its outgoing edges are unconditional edges. A non-source node is also a JOIN node. A JOIN node with multiple incoming edges is either an AND-JOIN node or an OR-JOIN node. All the incoming edges of an OR-JOIN node are mutually exclusive. A node is an AND-JOIN node if all its parent tasks are executed. We assume that each node with multiple incoming edges is either an OR-JOIN node or AND-JOIN node, and each node with multiple outgoing edges is either an OR-FORK node or an AND-FORK node. In the CTG shown in Fig. 1(a), v_3 is an OR-FORK node, v_1 an AND-FORK node, v_7 an OR-JOIN node, and v_3 an AND-JOIN node.

A scenario of a CTG is a sub-graph of the CTG formed by all the tasks in a complete execution trace of the task set. The probability $p(v_j)$ of a task v_j, is calculated as $p(v_j) = \Sigma_{s \in S_j} p(s)$, where $p(s) = \Pi_{c \in s} p(c)$ is the probability of the scenario s, and S_j is a set of all the scenarios in which v_j is executed and $p(c)$ is the probability of a condition c [24]. The MPSoC has a set $P = \{pe_1, pe_2, \cdots, pe_m\}$ of m processors.

We assume heterogeneous processors, where each processor $pe_k \in P$ is DVFS-enabled and can operate on a set $\{(V_{dd_1}, f_1), ..., (V_{dd_{n_k}}, f_{n_k})\}$ of n_k discrete voltage-frequency pairs. A matrix NC represents the execution times in clock cycles of all the tasks in G on different processors, where $NC(j, i)$ is the number of clock cycles of task v_i on pe_j.

The dynamic power $P_{d_{k,i}}$ of a task v_i on processor pe_k, dominated by discharging and charging of load capacitance due to gate switching, is given as $P_{d_{k,i}} = C_{eff_{k,i}} V_{dd_{k,i}}^2 f_{k,i}$ [3,7], where $C_{eff_{k,i}}$, $V_{dd_{k,i}}$ and $f_{k,i}$ are the effective load switching capacitance, the supply voltage and the operating frequency, respectively. The execution time of a task v_i on a processor pe_k operating at frequency $f_{k,i}$ is given as $t_{k,i} = NC(k,i)/f_{k,i}$. The operating frequency f is approximated by $f = ((1 + K_1)V_{dd} + K_2 V_{bs} - V_{th_1})^\alpha / K_6 L_d V_{dd}$ [3], where K_1, K_2, K_6 and V_{th_1} are circuit dependent constants, L_d the logic depth, and α the velocity saturation imposed by the used technology ($1.4 \leq \alpha < 2$). The total energy consumption $E_{k,i}$ of a task v_i on pe_k is computed as follows [3]:

$$E_{k,i} = NC(k,i)C_{eff_{k,i}} V_{dd_{k,i}}^2 + L_g(V_{dd_{k,i}} K_3 e^{K_4 V_{dd_{k,i}}} e^{K_5 V_{bs}} + |V_{bs}| I_j)t_{k,i} \quad (1)$$

We consider the 2D mesh NoC architecture, where each processor is associated with a router, and there are NR rows and NC columns. Every router has five ports with one port used to communicate with the associated processor, and the remaining four ports used to communicate with the neighboring routers. A link connecting two routers is called global link and a link connecting a router with its associated processor is referred to as local link. All the links are full duplex. All the global links are identical and have same link width (also called bus width or the number of wires) b_w.

We only take into account the energy consumption of global links and neglect the energy consumption of the local links. In the rest of this paper, all links refer to global links unless they are explicitly specified.

The NoC links can operate at a set $\{(V_{dd_1}, f_1), \cdots, (V_{dd_F}, f_F)\}$, of voltage frequency pairs. In a 2D mesh, the Manhattan distance $\eta_{i,j}$ between two processors pe_i and pe_j is defined as follows: $\eta_{i,j} = |x_i - x_j| + |y_i - y_j|$, where (x_i, y_i) and (x_j, y_j) are the coordinates of pe_i and pe_j, respectively.

The wormhole switching [22,25] and deterministic XY routing are used. We do not scale router frequencies as adjusting router frequencies makes the problem too complex. We assume that router frequencies are fixed and commensurate with link frequencies as in [22].

We transform the original CTG into an extended CTG so that communications can also be scheduled in the same way as tasks. The original CTG is transformed into an extended graph by adding additional nodes to G for every edge in original graph G. We refer to these additional nodes as communication nodes. The original nodes in graph G are kept unchanged and are referred to as task nodes. The extended graph is represented by $G_e(V + V^*, E', A')$, where V is the set of task node, V^* is the set of communication nodes, E' is the set of edges in the extended graph and A' is the set of triples where each element of the triple is an edge, the condition associated with the edge and probability of the condition. The extended graph G_e of CTG in Fig. 1(a) is shown in Fig. 1(b). In the rest of this paper, all the algorithms are based on the extended CTG.

Consider the message e_i for a communication node. The time taken by e_i on the links operating at frequency f_i such that e_i traverses the network without contention is calculated as follows [22]:

$$t_i = \frac{\chi_i}{b_w f_i} \tag{2}$$

We use the bit energy model given in [26,37] for communication. Assume that the source node and the destination node of e_i are mapped on processors pe_s and pe_d, respectively. The energy of transmitting one bit of the message e_i is $E_{bit} = (\eta_{s,d} + 1)E_{Rbit} + \eta_{s,d}E_{lbit_i}$, where E_{Rbit} is the energy consumption of one bit on one router, and E_{lbit_i} is the energy consumption of transmitting one bit on one link when all the links of e_i operate at f_i. Thus, the energy consumption of transmitting e_i on the links operating at frequency f_i is calculated as follows:

$$E_{comm_i} = \chi_i((\eta_{s,d} + 1)E_{Rbit} + \eta_{s,d}P_i/(f_i b_w)) \tag{3}$$

where P_i is the total power consumed in transmitting one bit when the links that e_i traverses operate at frequency f_i. P_i is the sum of the dynamic power P_{dyn_i} and static power P_{stat_i}, $P_i = P_{dyn_i} + P_{stat_i}$ [3]. The static and dynamic powers depend on how links are implemented. The frequency f_i is approximated by $f_i = ((1 + K_1)V_{dd} + K_2 V_{bs} - V_{th_1})^\alpha / K_6 L_d V_{dd}$ [3].

4 Task Scheduling and Frequency Assignment

4.1 Computing Successor-Tree-Consistent Deadlines

Our offline scheduling algorithm schedules nodes using the priorities of task nodes and communication nodes. We extend the notion of successor-tree-consistent deadline [35] to NoC-based MPSoCs, and propose a priority scheme for nodes, where the priority of each node v_i is its successor-tree-consistent deadline denoted by d'_i. When computing the successor-tree-consistent deadline of each node, we assume that all the processors and NoC-links operate at the maximum frequencies. Furthermore, the original CTG is used. Before defining the successor-tree-consistent deadline, we introduce the worst-case set of a task. Let $IPred(v_i)$ and $ISucc(v_i)$ be the sets of all the immediate predecessors and all the immediate successors of a task v_i, respectively.

Algorithm 1. Computing Successor-Tree-Consistent Deadlines

---

input : A conditional task graph $G(V, E, A)$ and a NOC-based MPSoC with
m processors

output: The successor-tree-consistent deadline d_i' of each task v_i in G

1   **for** *each node $v_i \in G$ in reverse topological order of G* **do**

2   **if** *outdegree(v_i) == 0* **then**

3       $d_i' = d_i$;

4       $WCS(v_i) = \emptyset$;

5   **else**

6   **if** *v_i is an OR-FORK node* **then**

7   Find a child v_j of v_i among all the children of v_i in the CTG such
         that $d_j' - \min_{\forall pe_k \in P}\{t_{k,s}\}$ is minimized;

8         $WCS(v_i) = \{v_j\} \cup WCS(v_j)$;

9         $d_i' = BackwardSchedule(v_i, WCS(v_i))$;

10   **else**

11         $WCS(v_i) = \emptyset$;

12   **for** *each $v_j \in ISucc(v_i)$* **do**

13           $WCS(v_i) = \{v_j\} \cup WCS(v_j) \cup WCS(v_i)$;

14         $d_i' = BackwardSchedule(v_i, WCS(v_i))$;

Definition 1. *The worst-case set of a task v_i, denoted by $WCS(v_i)$, is a set of tasks defined as follows:*

1. *If v_i is a sink node, $WCS(v_i) = \emptyset$.*
2. *If v_i is an OR-FORK node, $WCS(v_i) = \{v_j\} \cup WCS(v_j)$, where v_j is in $ISucc(v_i)$ and satisfies $d_j' - \min_{\forall pe_k \in P}\{t_{k,i}\} = \min_{\forall v_s \in ISucc(v_i)}\{d_s' - \min_{\forall pe_k \in P}\{t_{k,s}\}\}$.*
3. *If v_i is an AND-FORK node, $WCS(v_i) = \bigcup_{v_s \in ISucc(v_i)} (WCS(v_s) \cup \{v_s\})$.*

Definition 2. *Given a CTG G and a task v_i, the successor tree of a task v_i is a weighted directed tree $ST(G,v_i) = (V', E', X')$ where $v' = \{v_i\} \cup WCS(v_i)$, $E' = \{(v_i, v_j) : v_j \in WCS(v_i)\}$ and $X' = \{\chi_s' :$ if v_j is the immediate successor of v_i, $\chi_s' = \chi_s$, the edge weight of $(v_i, v_j) \in E$; otherwise, $\chi_s' = 0\}$.*

Definition 3. *Given a task v_i, if v_i is a sink task, its successor-tree-consistent deadline d_i' is equal to its preassigned deadline d_i. Otherwise, d_i' is the upper bound on the latest completion time of v_i in any feasible schedule of the relaxed problem instance: a set $V' = \{v_i\} \cup WCS(v_i)$ of tasks with the precedence constraints in the form of the weighted successor tree $ST(G, v_i)$, where the deadline of each task $v_j \in WCS(v_i)$ is its successor-tree-consistent deadline, and the deadline of v_i is its preassigned deadline, and the same MPSoC.*

Algorithm 1 describes our successor-tree-consistent deadline algorithm. We select each node v_i in CTG G in reverse topological order of G (Line 1). If the

---

Algorithm 2. Computing Backward Schedule

---

1 **Function** $BackwardSchedule(v_i, WCS(v_i))$
2   Partition the tasks in $WCS(v_i)$ into two disjoint sets U and J such that U consists of all the tasks in $WCS(v_i)$ each of which does not receive any data from v_i, and J contains all the tasks in $WCS(v_i)$ that are not in U;
3   Sort all the tasks in U in non-increasing order of their successor-tree-consistent deadlines;
4   Schedule each task v_j in U on a processor that maximizes its start time;
5   Sort all the tasks in J in non-increasing order of their successor-tree-consistent deadlines and for the tasks with the same successor-tree-consistent deadlines, sort them in non-increasing order of their edge weights;
6   Schedule each task v_j in J on a processor that maximizes its start time;
7   Schedule v_i on a processor that maximizes its completion time respecting the constraints specified by the successor tree of v_i;
8   Set d'_i to the completion time of v_i;
9   **return** d'_i;

---

node v_i is a sink node, its successor-tree-consistent deadline is equal to its pre-assigned deadline and its worst case set is an empty set (Lines 2–4). If v_i is not a sink and is an OR-FORK node, we find a child v_j of v_i among all the children of v_i in the CTG such that $d'_j - \min_{\forall pe_k \in P}\{t_{k,j}\}$ is minimized (Lines 6–7). In this case the worst-case set of v_i is the union of node v_j and the worst-case set of v_j. Given the worst-case set of v_i we calculate its successor-tree-consistent deadline using Algorithm 2. If v_i is neither a sink node nor an OR-FORK node, we first calculate the worst-case set of v_i (Lines 11–13), and then calculate its successor-tree-consistent deadline by calling Algorithm 2 (Line 14).

4.2 Earliest Successor-Tree-Consistent Deadline First Algorithm

The number of scenarios in a CTG grows exponentially as the number of conditions increases. Therefore, our offline scheduling approach constructs a single unified schedule for all the scenarios by exploiting the mutual exclusion relations between communication and task nodes. In a CTG, two nodes are said to be concurrent if they are not reachable from each other and are not mutually exclusive.

We propose an Earliest-Successor-Tree-consistent Deadline First (ESTDF) scheduling algorithm assuming that all processors and links operate at the maximum frequencies. Algorithm 3 describes the ESTDF algorithm. ESTDF is called by our main algorithm IOETCS described in the next subsection. It determines the order in which task nodes and communication nodes are executed and captures this order by adding additional precedence constraints in the input graph G. Schedule constructed by prioritizing nodes with shorter successor-tree-deadline over nodes with longer successor-tree-deadline enables DVFS algorithm to save more energy as it can efficiently utilize slack available for the nodes.

Algorithm 3. Earliest Successor-Tree-Consistent Deadline First Scheduling

---

    input : CTG G, a matrix NC of worst-case clock cycles of tasks, a vector X of communication volumes and a task-to-processor mapping Map

    output: Graph G that captures the precedence as well the additional resource constraints

1 Construct a list R of all the source nodes in G;

2 **while** *R is not empty* **do**

3   Find a node $v_i \in R$ with minimum successor-tree-consistent deadline and compute its ready time r_i;

4   **if** *v_i is a communication node* **then**

5         $\xi_i = r_i + t_i$;

6   Insert unconditional directed edges in G from v_i to the communication nodes satisfying conditions C1, C2 and C3;

7   **else**

8         $\xi_i = r_i + t_{k,i}$, where $t_{k,i}$ is the execution time of v_i on pe_k at maximum frequency;

9   Insert unconditional directed edges in G from v_i to unscheduled nodes satisfying conditions C4, C5 and C6;

10   Insert all the ready nodes in R;

Therefore, we first construct a list R of all the source nodes in G (Line 1). Next, we select a node from R with the highest priority, schedule it and insert in R all the ready nodes. Ready nodes are those nodes whose parents have been scheduled. This process is repeated until all the nodes in G has been scheduled (Lines 2–10).

A node has the highest priority if it has the shortest successor-tree-consistent deadline among all the nodes in R (Line 3). Thus, nodes with shorter successor-tree-consistent deadlines are scheduled earlier than nodes with longer successor-tree-consistent deadlines. Each time a node is selected for scheduling, we calculate its ready time (Line 3). The ready time of a node v_i is the earliest time it can start its execution. A node cannot start its execution until all its predecessors have finished execution. Thus the ready time of node v_j is:

$$r_j = \max\{\xi_l : v_l \in IPred(v_j)\}, \tag{4}$$

where ξ_l is the finish time of node v_l. If v_i is communication node, we compute its finish time and insert additional edges in G called resource constraints from v_i to the nodes that satisfy the following three conditions simultaneously (Lines 4–6).

– C1: They are concurrent to v_j. Two communication nodes are said to be concurrent if they are not reachable in CTG G from each other and are not mutually exclusive. Concurrent nodes can start their executions at the same time if allocated to the same resource since there are no precedence constraints between them. This constraint together with C2 and C3 enforce a total order for communication nodes.

- C2: Their successor-tree-consistent deadline is either longer than or equal to v_j.
- C3: They traverse the same communication links that v_j traverses. Resource constraints are only introduced between communication nodes that compete for the same communication links.

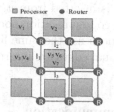

(a) Task-to-processor mapping

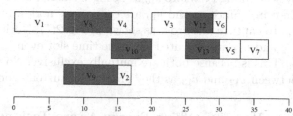

(b) Local schedule constructed in ESTDF manner

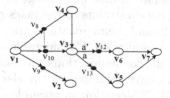

(c) Graph capturing the precedence and resource constraint

Fig. 2. An illustrative example

These constraints are enforced so that no two non-mutually exclusive communication nodes are allocated the same resource at the same time. If v_i is a task node, we calculate its finish time in a similar way and insert resource constraints from v_i to the nodes in G that simultaneously satisfy following three conditions:

- C4: They are concurrent to v_j.
- C5: They are assigned to the same processor where v_j is allocated.
- C6: They have not been scheduled.

Consider the CTG in Fig. 1(b) and the MPSoC in Fig. 2(a) where all the processors are identical. The execution times of tasks at the maximum processor frequency are $t_{1,1} = 7$, $t_{1,2} = 2$, $t_{1,3} = 5$, $t_{1,4} = 3$, $t_{1,5} = 3$, $t_{1,6} = 2$, and $t_{1,7} = 4$ time units. The communication times are $t_8 = 7$, $t_9 = 8$, $t_{10} = 6$, $t_{11} = 5$, $t_{12} = 4$, $t_{13} = 5$, $t_{14} = 7$, and $t_{15} = 9$ time units. All the tasks have a common deadline of 40 time units. Consider the task mapping in Fig. 2(a). Based on this task mapping, the input CTG G_e shown in Fig. 1(b) to ESTDF does not contain communication nodes v_{11}, v_{14} and v_{15}. Furthermore, edges (v_4, v_{11}), (v_{11}, v_3), (v_6, v_{14}), (v_{14}, v_7), (v_5, v_{15}), (v_{15}, v_7) in G_e are replaced by (v_4, v_3), (v_6, v_7) and (v_5, v_7). Figure 2(b) gives an illustration of ESTDF scheduling algorithm for task mapping in Fig. 2(a). Three communication nodes v_8, v_{10} and v_9 become ready after v_1 is scheduled. Communication nodes v_8 and v_{10} traverse the same link l_1. Since they are concurrent, they contend for l_1. ESTDF resolves this conflict by scheduling v_8 before v_{10} as v_8 has a smaller successor-tree-consistent deadline than v_{10}. Since v_8 and v_{10} are concurrent nodes, an edge is inserted from v_8 to v_{10} to capture this order as shown in Fig. 2(c). Notice that communication nodes v_{12} and v_{13} are allocated the same time slot even though both use the same link l_3. This is because both are mutually exclusive. No additional edges are inserted between v_{12} and v_{13} as they are not concurrent nodes.

4.3   Iterative Offline Energy-Aware Task and Communication Scheduling Algorithm (IOETCS)

A heterogeneous MPSoC consist of processors that have different power performance profiles and may operate at different voltage and frequency levels. Moreover, deadline and precedence constraints of the tasks must be observed. Therefore, the order in which tasks and communications are executed may significantly impact the overall energy consumption. Considerable amount of energy may be saved by prioritizing tasks and communications with shorter deadlines over tasks and communications with longer deadlines as the slack available for tasks can be efficiently utilized by DVFS algorithm to assign low voltages and frequencies to them. Consequently, the quality of a solution obtained by an energy-efficient scheduling algorithm is influenced by three factors: task mapping, ordering and voltage assignment. The state-of-the-art approach [22] performs task ordering and voltage scaling in an integrated manner and performs task mapping separately. Our observation is that task and communication ordering and voltage scaling are helpful in steering the task mapping optimization process towards a more energy-efficient solution. This is one of major factors that we consider in design of our Iterative Offline Energy-aware Task and Communication Scheduling Algorithm (IOETCS).

In our IOETCS algorithm we perform mapping, scheduling and voltage scaling in an integrated manner. We schedule nodes by first calculating their successor-tree-consistent deadlines (Line 1). We select each node from list L in order and tentatively assign it to each processor (Lines 3–5). Each time we assign (tentatively) a node in L to a processor we perform four major steps. First, we construct a subgraph $G_s(V_s + V_s^*, E_s)$ (Lines 7–8). In subgraph G_s, V_s is the set of all the mapped task nodes, V_s^* the set of communication nodes with both child and parent nodes mapped on different processors and E_s the set of all the edges where every edge in E_s belongs to E' and both its head and tail nodes are in $V_s + V_s^*$. For each communication node v_s whose parent node v_p and child node v_c are mapped on the same processor, insert a directed edge (v_p, v_c) to E_s. Second, given the subgraph G_s, we call $ESTDF()$ $(G_s = ESTDF(G_s, NC, X, Map))$ to construct a local schedule and capture the resource constraints introduced by the local schedule (Line 9). Third, given a task-to-processor mapping and a graph G_s, assign voltages/frequencies to task and communication nodes by solving an NLP problem (Line 9). The objective of the NLP problem is to minimize the total expected energy consumption of graph G_s. The expected energy consumption is given as $E_{exp} = \Sigma_{v_i \in V_s} p(v_i) E_{k,i} + \Sigma_{v_i \in V_s^*} p(v_i) E_{comm_i}$. The NLP problem is formulated as follows:

minimize E_{exp}

s.t.

$$t_{k,i} = \frac{NC_{k,i} K_6 L_d V_{dd_{k,i}}}{((1 + K_1)V_{dd_{k,i}} + K_2 V_{bs} - V_{th_1})^\alpha} \qquad \forall v_i \in V_s \qquad (5)$$

$$t_i = \frac{\chi_i K_6 L_d V_{dd_i}}{b_w((1 + K_1)V_{dd_i} + K_2 V_{bs} - V_{th_1})^\alpha} \qquad \forall v_i \in V_s^* \qquad (6)$$

$$\rho_i + t_{k,i} \leq d_i' \qquad v_i \in V_s \qquad (7)$$

$$\rho_i + t_{k,i} \leq \rho_j \qquad \forall(v_i, v_j) \in E_s \wedge v_i \in V_s \qquad (8)$$

$$\rho_i + t_i \leq \rho_j \qquad \forall(v_i, v_j) \in E \wedge v_i \in V_s^* \qquad (9)$$

$$\rho_i \geq 0 \qquad \forall v_i \in V_s + V_s^* \qquad (10)$$

$$V_{dd_{k,min}} \leq V_{dd_{k,i}} \leq V_{dd_{k,max}} \qquad (11)$$

$$V_{dd_{min}} \leq V_{dd_i} \leq V_{dd_{max}} \qquad \forall v_i \in V^* \qquad (12)$$

The decision variables are, the start time ρ_i, the task node execution time $t_{k,i}$, the communication time t_i, the task voltage $V_{dd_{k,i}}$ and the communication voltage V_{dd_i}. $V_{dd_{k,min}}$ and $V_{dd_{k,max}}$ are the minimum and the maximum supply voltages of the processor pe_k, respectively. Equations (5) and (6) are the task execution time and communication time constraints, respectively. Equation (7) is the deadline constraint, and Eqs. (8) and (9) are the precedence constraints. Since the constraints and the objective function are convex, this NLP problem can be solved in polynomial time [5, 38].

Next, assign each node in G_s a valid discrete frequency given a frequency assigned to it by an NLP and compute the total expected energy consumption E_{exp} (Line 11). Note that in our NLP formulation the decision variables for supply voltages are continuous. However, typically processors and communication links can only operate at a set of fixed voltage and frequency levels. Under the continuous supply voltage constraints the task and communication nodes may be assigned invalid frequencies by the NLP. Hence, we need to assign them valid discrete frequencies.

5   Discrete Frequency Assignment Algorithms

In this section we describe our ILP-based algorithm and heuristic algorithm to assign each node a valid discrete frequency.

Algorithm 4. IOETCS

---

input : CTG $G_e(V + V^*, E', A')$ with a matrix NC and a set X, node
        deadlines, and a NoC-based MPSoC
output: Schedule graph $G^*(V_s + V_s^*, E_s)$, a vector Map for task mapping,
        and a communication and task voltage assignment

1 Construct a list L of nodes in V sorted in non-descending order of
  successor-tree-consistent deadlines;
2 $\forall v_i \in V\,Map[i] = 0$;
3 **for** *each $v_i \in L$ in order* **do**
4 | $E_{ini} = \infty$;
5 | $p = 0$;

6 **for** *each $pe_k \in P$* **do**
7 | $Map[i] = k$;
8 | Construct graph G_s;
9 | $G_s = ESTDF(G_s, NC, X, Map)$;
10 | Compute the voltage assignment of nodes in G_s and the total expected
           energy E_{exp} of G_s by solving NLP;
11 | Assign nodes in G_s valid discrete frequencies and compute the total
           expected energy E_{exp} using the discrete frequency assignment algorithm;
12 | **if** $E_{exp} < E_{ini}$ **then**
13 | | $G^* = G_s$;
14 | | $E_{ini} = E_{exp}$;
15 | | $p = k$;

16 $Map[i] = p$;

5.1 ILP-Based Algorithm

The optimal frequency f_i^{opt} of a communication node and the optimal frequency $f_{k,i}^{opt}$ of a task node are computed as described in Sect. 4. We differentiate between the following two cases for each task or communication node v_i:

1. If v_i is a task node and its frequency $f_{k,i}^{opt}$ is a discrete frequency of the processor pe_k where v_i is assigned, assign $f_{k,i}^{opt}$ to v_i. If v_i is a communication node and its frequency f_i^{opt} is equal to a discrete link frequency, assign f_i^{opt} to v_i.

2. If v_i is a task node and its frequency $f_{k,i}^{opt}$ is not a discrete frequency of the processor pe_k where v_i is assigned, find two frequencies $f_{k,i}^{opt,u}$ and $f_{k,i}^{opt,l}$ of the pe_k where v_i is assigned such that $f_{k,i}^{opt,u}$ is the smallest discrete frequency of pe_k larger than $f_{k,i}^{opt}$ and $f_{k,i}^{opt,l}$ is the largest discrete frequency of pe_k smaller than $f_{k,i}^{opt}$. Similarly, if v_i is a communication node and its frequency f_i^{opt} is not a discrete link frequency, find two discrete frequencies $f_i^{opt,l}$ and $f_i^{opt,u}$ of communication links such that $f_i^{opt,u}$ is the smallest discrete frequency of communication links larger than f_i^{opt} and $f_i^{opt,l}$ is the largest discrete frequency of communication links smaller than f_i^{opt}. We introduce a binary

decision variable to select between $f_i^{opt,u}$ and $f_i^{opt,l}$ if v_i is a communication node or between $f_{k,i}^{opt,u}$ and $f_{k,i}^{opt,l}$ if v_i is a task node.

We introduce a binary decision variable to select between $f_i^{opt,u}$ and $f_i^{opt,l}$ if v_i is a communication node or between $f_{k,i}^{opt,u}$ and $f_{k,i}^{opt,l}$ if v_i is a task node:

$$x_i = \begin{cases} 0 & \text{if } v_i \text{ uses } f_i^{opt,l} \text{ or } f_{k,i}^{opt,l} \\ 1 & \text{if } v_i \text{ uses } f_i^{opt,u} \text{ or } f_{k,i}^{opt,u} \end{cases}$$

Let V^{opt} be a set of nodes that lie in Case 1. $V_R = V_s - V^{opt}$ is a set of task nodes and $V_R^* = V_s^* - V^{opt}$ is a set of communication nodes for which Case 2 holds. The expected energy consumption is now given as $E_{exp} = \sum_{v_i \in V_R}((1 - x_i)E_{k,i}^{opt,l} + x_i E_{k,i}^{opt,u})p(v_i) + \sum_{v_i \in V_R^*}((1 - x_i)E_{comm_i}^{opt,l} + x_i E_{comm_i}^{opt,u})p(v_i) + C$, where $E_{k,i}^{opt,l}$ and $E_{k,i}^{opt,u}$ (given in Eq. (1)) are the energy consumptions of a task node v_i on a processor pe_k at the frequencies $f_{k,i}^{opt,l}$ and $f_{k,i}^{opt,u}$, respectively, $E_{comm_i}^{opt,l}$ and $E_{comm_i}^{opt,u}$ (given in Eq. (3)) are the energy consumptions of a communication node v_i when all the links on its routing path operate at the frequencies $f_i^{opt,l}$ and $f_i^{opt,u}$, respectively, and C is the sum of energy consumption of nodes in V^{opt}. The ILP problem is formulated as follows:

minimize E_{exp}

s.t.

$$t_{k,i} = t_{k,i}^{opt,l}(1 - x_i) + t_{k,i}^{opt,u} x_i \qquad\qquad \forall v_i \in V_R \qquad (13)$$

$$t_i = t_i^{opt,l}(1 - x_i) + t_i^{opt,u} x_i \qquad\qquad \forall v_i \in V_R^* \qquad (14)$$

$$\rho_i + t_{k,i} \le d_i' \qquad\qquad \forall v_i \in V_R \qquad (15)$$

$$\rho_i + t_{k,i}^{opt} \le d_i' \qquad\qquad \forall v_i \in V^{opt} \cup V_s \qquad (16)$$

$$\rho(v_i) + t_{k,i}^{opt} \le \rho_j \qquad\qquad \forall (v_i, v_j) \in E_s \wedge v_i \in V^{opt} \cup V_s \qquad (17)$$

$$\rho(v_i) + t_i^{opt} \le \rho_j \qquad\qquad \forall (v_i, v_j) \in E_s \wedge v_i \in V^{opt} \cup V_s^* \qquad (18)$$

$$\rho(v_i) + t_{k,i} \le \rho_j \qquad\qquad \forall (v_i, v_j) \in E_s \wedge v_i \in V_R \qquad (19)$$

$$\rho(v_i) + t_i \le \rho_j \qquad\qquad \forall (v_i, v_j) \in E_s \wedge v_i \in V_R^* \qquad (20)$$

$$\rho_i \ge 0 \qquad\qquad \forall v_i \in V_s \cup V_s^* \qquad (21)$$

The decision variables are task execution time $t_{k,i}$, communication time t_i, binary variable x_i and start time ρ_i. $t_{k,i}^{opt,l}$ and $t_{k,i}^{opt,u}$ are the execution times of the task node v_i on the processor pe_k where v_i is mapped at the frequencies $f_{k,i}^{opt,l}$ and $f_{k,i}^{opt,u}$, respectively. $t_i^{opt,l}$ and $t_i^{opt,u}$ are the communication times (given in Eq. (2)) of the communication node v_i when all the links of the communication path operate at the frequencies $f_i^{opt,l}$ and $f_i^{opt,u}$, respectively. $t_{k,i}^{opt}$ is the execution time of the task node v_i at frequency level $f_{k,i}^{opt}$ and t_i^{opt} is the communication time of the communication node v_i when all the links of the communication operate at the frequency f_i^{opt}. Equation (13) defines the execution time of a task node. Equation (14) defines the communication time of a communication node. Equations (15) and (16) collectively define the deadline constraints, Eqs. (17), (18), (19) and (20) collectively define the precedence constraints.

In our ILP formulation the binary variable x_i determines if a node v_i executes at $f_{k,i}^{opt,u}$ (or $f_i^{opt,u}$) or at $f_{k,i}^{opt,l}$ (or $f_i^{opt,l}$). Equations (13) and (14) in our ILP formulation implement this OR logic. Since x_i is a binary decision variable it can either be 1 or 0. When x_i is 0, Eq. (13) becomes $t_{k,i} = t_{k,i}^{opt,l}(1-0)+0t_{k,i}^{opt,u} = t_{k,i}^{opt,l}$ and thus, the task executes at $f_{k,i}^{opt,l}$ and the energy consumed by v_i is reflected by the term $(E_{k,i}^{opt,l}(1 - x_i) + E_{k,i}^{opt,u}x_i)p(v_i) = (E_{k,i}^{opt,l}(1 - 0) + E_{k,i}^{opt,u}0)p(v_i) = E_{k,i}^{opt,l}p(v_i))$ in the objective function. Similarly, when x_i equals 1 the first term in Eq. (13) becomes 0 and this represents the scenario when v_i operates at $f_{k,i}^{opt,u}$. Notice that when the execution time changes depending on the value of decision variable x_i, the start time of other nodes in the graph may also change. This is because of the precedence and resource constraints between the nodes and therefore, we have decision variable ρ_i to capture the change in start time.

5.2 Heuristic Algorithm

The ILP problem is NP-complete [11]. Therefore, the previous ILP-based algorithm is not scalable. Next, we propose a polynomial time heuristic to assign discrete frequencies to task and communication nodes. Algorithm 5 describes our discrete frequency and voltage assignment heuristic algorithm. In our heuristic we use the schedule constructed by IOETCS algorithm (Algorithm 4). We first compute all the node cuts of graph G^* (Line 1). The node cuts $C_p(p = 1, 2, 3, \cdots)$ of graph G^* are computed as follows:

– Create a copy G' of G^* and repeat the following steps until G' is empty:
  1. Create an empty cut C_p.
  2. Add all the the source nodes with zero in-degree in G' to C_p.
  3. Remove all the source nodes and their incident edges from G'.

The NLP algorithm assigns task and communication nodes frequencies in a continuous range. Thus, the frequency assigned by NLP may not be a valid discrete frequency. Therefore, we inspect the frequencies assigned to nodes by NLP (Lines 2–6). If the frequency assigned to a task or a communication node v_i is a valid discrete frequency, it is not changed (Line 4). Otherwise v_i is assigned a nearest valid discrete frequency lower than the assigned frequency (Line 6). After assigning the nodes valid discrete frequency we construct a new local schedule using the new frequency such that the order between nodes remains the same as in the schedule used by the NLP-based algorithm (Line 7). We construct a new schedule because under the new frequency assignment some tasks may miss their deadlines. If the new schedule is feasible, the algorithm terminates. Otherwise, we find a minimum set of nodes such that by increasing the frequency of each

---

Algorithm 5. Discrete Frequency and Voltage Assignment Heuristic

---

input : Schedule graph G^* , Vector map, and a NoC-based MPSoC
output: A schedule that assigns to each task and communication a valid
          discrete frequency, start time and an execution time

1 Compute the node cuts of graph G^*;
2 **for** *each $v_i \in V_s + V_s^*$* **do**
3 | **if** *f_i^{opt} is a valid discrete frequency* **then**
4 | | Assign f_i^{opt} to v_i;
5 | **else**
6 | | Assign $f_{k,i}^{opt,l}$ to v_i if v_i is a task node or $f_i^{opt,l}$ to v_i if v_i is a
   | | communication node;

7 Construct a new local schedule using the new frequency such that the order
   between nodes remains the same as in the schedule used by the NLP-based
   algorithm;
8 **while** *there is a late task node* **do**
9 | Find the first late task node v_j;
10 | Find a set B of nodes blocking v_j;
11 | **while** *v_j is late* **do**
12 | | Compute the rank of each node $v_i \in B$;
13 | | Select a node v_i with the highest rank by comparing ranks
   | | lexicographically;
14 | | Adjust the frequency of v_i to $f_i^{opt,u}$ if v_i is a communication node or
   | | $f_{k,i}^{opt,u}$ if v_i is a task node;
15 | | Remove v_i from set B;
16 | | Update the schedule;

---

node v_i in this set (to $f_i^{opt,u}$ if v_i is a communication node or $f_{k,i}^{opt,u}$ if v_i is a task node), makes the schedule feasible and there is a minimal increase in the energy consumption.

Next, we describe a method for finding the set of these nodes (Lines 8–15). We scan the schedule in non-decreasing order of time to find the first late task node v_j (Line 9). For this node, we find a set B of blocking nodes (Line 11). Each node $v_z \in B$ satisfies the following two conditions:

1. v_z belongs to the set $\{v_j\} \cup Pred(v_j)$, where $Pred(v_j)$ is a set of predecessors of v_j.
2. The frequency of v_z can be adjusted. The frequency of v_z can be adjusted if it is not a valid discrete frequency assigned by NLP and is not $f_{k,z}^{opt,u}$ if v_z is a task node or $f_z^{opt,u}$ if v_z is a communication node.

The set B consists of nodes that affect the finish time of the late task node v_j. Therefore, increasing the frequency of some or all the nodes in B may result in an early start of v_j and consequently allow v_j to meet its deadline. We compute the ranks of all the nodes in set B to find a node v_h with the highest rank by

Table 1. Execution times of tasks in CTG shown in Fig. 3(a) at maximum processor frequency

Task	Type 1	Type 2	Task	Type 1	Type 2	Task	Type 1	Type 2
v_1	2.9	1.3	v_3	3.12	1.7	v_5	2.32	1.12
v_2	2.2	1.9	v_4	2.46	1.9	v_6	3.55	1.86
v_7	2.4	1.2	v_9	2.1	1.9	v_{11}	2.59	1.5
v_8	2.7	1.7	v_{10}	3.1	1.7			

Table 2. Processor configurations

	Type 1				Type 2			
Supply voltage (V)	0.75	0.70	0.65	0.60	0.95	0.90	0.85	0.80
Frequency (GHz)	1.53	1.26	1.01	0.78	2.7	2.4	2.1	1.8

comparing the ranks lexicographically (Line 13). The rank of each node $v_i \in B$ is a 2-tuple (g_i, κ_i) which reflects the impact of v_i on shifting the late node v_j to an earlier time. Let C_p be a set of nodes of a cut containing v_i, C_p' be $C_p \cap B$, FT_j^{old} the finish time of v_j in the current schedule, FT_j^{new} the finish time of v_j after the frequencies of all the nodes in the set C_p' are increased by one level, and $FT_j^{new,i}$ the finish time of v_j when the frequency of v_i is increased by one level. The normalized time gain g_i of the cut containing v_i, is given as

$$g_i = \frac{FT_j^{old} - FT_j^{new}}{E_{exp}' - E_{exp}} \tag{22}$$

where E_{exp}' is the expected energy consumption after the frequencies of all the nodes in the set C_p' are increased by one level. The normalized time gain κ_i of v_i is computed as follows:

$$\kappa_i = \frac{FT_j^{old} - FT_j^{new,i}}{E_{exp}'' - E_{exp}} \tag{23}$$

where E_{exp}'' is the expected energy consumption after the frequency of v_i is increased by one level.

We increase the frequency of v_h and update the schedule (Lines 14–16). This process of finding a blocking nodes of late task with highest rank and increasing its frequency is repeated until the late task node meets its deadline.

The process of scanning the schedule to find the late task node to pushing its start time to an earlier time until the node meets its deadline is repeated until there is no late task node in the schedule.

Figures 4, 5 and 6 show an illustrative example of our discrete frequency assignment heuristic where task execution times at the maximum processor frequency are shown in Table 1. Figure 4(a) shows the task-to-processor mapping generated by IOETCS algorithm for the example shown in Fig. 3, and Fig. 4(b) shows the

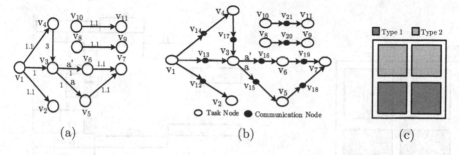

Fig. 3. (a) CTG G with edge weights are communication volumes in mega bits (b) Extended Graph G_e (c) 2×2 NoC-based MPSoC with two types of processors

Table 3. NoC link configurations

Supply voltage (V)	0.55	0.50	0.43
Frequency (MHz)	600	400	200
Bandwidth (Gbps)	4.7	3.1	1.55

schedule generated under the frequencies assigned by the NLP. Tables 2 and 3 show the discrete frequencies for processors and communication links. Notice that as shown in Table 4 task nodes v_1, v_3, v_4, v_8, v_9 and v_{10} are assigned invalid frequencies by NLP (frequencies assigned to nodes are shown on top of each node schedule in all schedules). This is because NLP assumes the processor can operate at any frequency in a continuous range. Our discrete frequency assignment heuristic assigns them valid discrete frequencies.

We first find the cuts of the graph. The cuts of the graph are shown in Fig. 5. After this, we analyze the frequencies assigned to the nodes by the NLP. If the frequency assigned to a node is valid discrete frequency, it is not changed. Otherwise, we assign this node a lower frequency ($f_{k,i}^{opt,l}$ or $f_i^{opt,l}$). Figure 5(b) shows the schedule when nodes v_1, v_3, v_4, v_8, v_9 and v_{10} operate at nearest valid lower discrete frequency assigned to them by NLP ($f_{k,i}^{opt,l}$). Under these frequencies, the schedule is not feasible. The late task nodes are highlighted in Fig. 5(b).

Next, we scan the schedule in non-decreasing order of time to find the first late task node and this node is highlighted in Fig. 5(b). The first late task node is v_2. We find a set B of its blocking nodes. These nodes are shown in Fig. 5(c). Blocking

Table 4. Nodes assigned invalid frequencies by NLP

Node	Lower frequency $f_{k,i}^{opt,l}/f_i^{opt,l}$	NLP assigned frequency f_i^{opt}	Upper frequency $f_{k,i}^{opt,u}/f_i^{opt,u}$
v_1 v_3 v_4	1.26	1.47	1.53
v_8 v_9 v_{10}	0.78	0.87	1.01

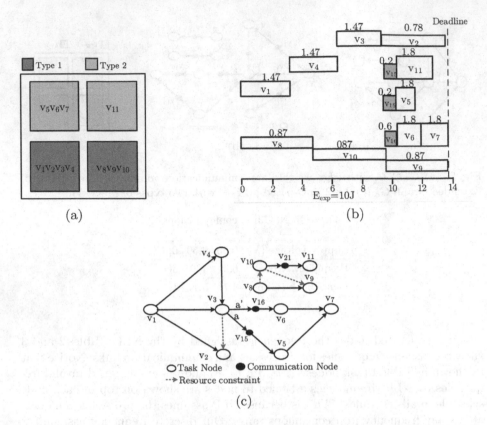

Fig. 4. (a) Task-to-processor mapping generated by IOETCS algorithm (b) Schedule generated by frequencies assigned to tasks by NLP and the integer on top of each node is the frequency in GHz assigned to the node (c) CTG with resource constraints

nodes satisfy two criteria. Firstly, they are predecessors of v_2. Secondly, their frequencies can be adjusted. The frequency of a node v_i cannot be adjusted if it has been assigned a valid discrete frequency by NLP or it has already been assigned $f_{k,i}^{opt,u}$ or $f_i^{opt,u}$. Now, we calculate the rank of each blocking node. The rank of each blocking node is a two-tuple. The first element of the tuple is normalized time gain of the node cut containing the node and the second element is the normalized time gain of the node. Node v_2 has three blocking nodes: v_1, v_3 and v_4. To calculate the rank of v_3, we first calculate the normalize node cut gain of the cut containing v_3. The node cut of v_3 is C_3 as shown in Fig. 5(a). Next, we find the nodes in C_3 that are also in B. The only common node in both C_3 and B is v_3. We increase the frequency of v_3 to $f_{3,3}^{opt,u} = 1.5\,\text{GHz}$ and generate a new schedule such that the order between the tasks remains the same as in the schedule by NLP. Given the new finish time of the v_2 and the expected energy consumption, the normalized time gain of C_3 is 4.331. Similarly, we compute the normalized time gain of v_3 which is 4.331. Thus, the rank of v_3 is $(4.331, 4.331)$. The ranks of v_1 and v_4 are calculated

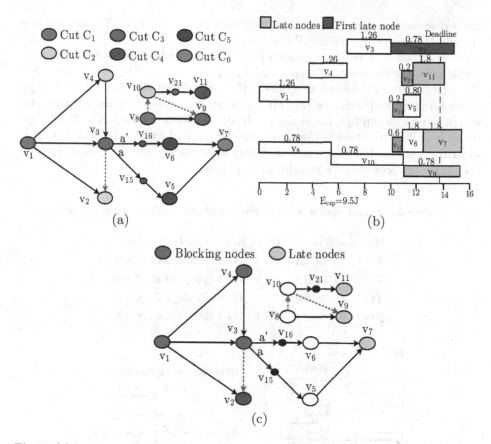

Fig. 5. (a) Node cuts (b) Schedule when a set of nodes assigned invalid frequencies by NLP are executed at a lower frequency $f_{k,i}^{opt,l}$ (c) Blocking nodes of the first late task node

next which are $(4.318, 4.318)$ and $(4.330, 4.330)$, respectively. By comparing their ranks lexicographically the node v_3 has the highest rank. Its frequency is updated to $f_{3,3}^{opt,u} = 1.5$ GHz. Figure 6(a) shows the new schedule under the new frequency of v_3. We next delete v_3 from B and check if v_2 is late. Since v_2 is still late, we repeat the whole process again to find a node with the highest rank in B. Now v_4 has the highest rank. Similarly, the frequency of v_4 is adjusted and a new schedule with the new frequency of v_4 is constructed as shown in Fig. 6(c). Notice that in the schedule shown in Fig. 6(c) not only v_2 but also v_7 meets its deadline.

After that, we select the next late task which is v_8 and repeat the same process until all the nodes meet their deadlines.

6   Performance Evaluation

In this section, we use IOETCS-ILP and IOETCS-Heuristic to denote our approach using the ILP-based algorithm and the heuristic, respectively, for assigning a discrete frequency to each task and each communication. To demonstrate the effectiveness of IOETCS-ILP and IOETCS-Heuristic, we compare them with three approaches. The first approach is Li-Wu approach, a state-of-the-art approach for unconditional task graph model proposed in [22]. The second approach ILP-vpv-flv that is the same as IOETCS-ILP except that the NLP and ILP algorithms are modified such that they only scale processors frequencies/voltages and assign the maximum link frequency to all communication nodes. The third approach is ILP-fpv-vlv that is the same as IOETCS-ILP except the NLP and ILP algorithms are modified such that they only scale the voltages of links and assign the maximum processor frequencies to task nodes.

Table 5. Characteristics of benchmarks without conditional precedence

BM	$a/b/D$	Dim	BM	$a/b/D$	Dim
TG 1	17/19/1.4	4×5	TG 2	20/24/0.77	5×4
TG 2	15/11/0.98	4×5	TG 4	16/12/0.89	5×4
TG 5	27/28/2.4	6×5	TG 6	27/35/2.7	6×5
TG 7	27/39/2.9	6×5	TG 8	30/40/3.45	6×6

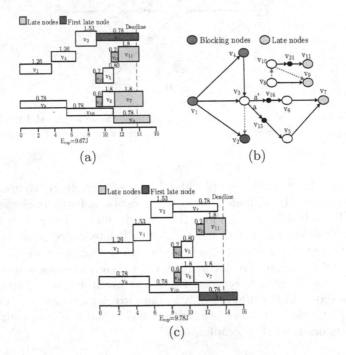

Fig. 6. (a) Schedule when v_3 operates at higher frequency (b) Blocking nodes of v_2 (c) Schedule when v_3 and v_4 operate at a higher frequency

Table 6. Characteristics of benchmarks with conditional precedence constraints.

BM	$x/y/z/D$	Dim	BM	$x/y/z/D$	Dim
CTG 1	17/2/6/0.74	3×3	CTG 2	20/1/2/1.06	3×3
CTG 3	15/2/4/0.723	3×3	CTG 4	17/2/6/0.93	3×3
CTG 5	30/4/11/1.73	3×3	CTG 6	35/3/8/3.101	3×2
CTG 7	33/5/15/3.7128	3×2	CTG 8	31/3/9/3.69	3×2

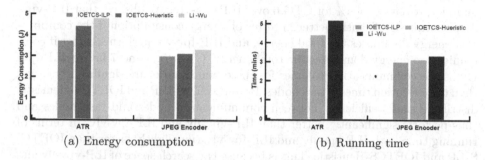

(a) Energy consumption (b) Running time

Fig. 7. Comparison of real-world benchmarks without conditional precedence constraints

6.1 Experimental Setup

We use the same experimental setup as in [3, 9, 13]. The technology parameters are taken from [9]. We use two types of processors in our experiments, Type 1 and Type 2, modelled after the processors in [9] and [10], respectively. The configuration for NoC links is adopted from [22]. The execution times in cycles of tasks are randomly generated within $[10, 100] \times 106$ and $[5, 10] \times 106$, respectively. The communication volumes are generated randomly within $[80, 800] \times 106$ in bits. The deadline for each application is set to twice the makespan of the schedule of the application constructed by IOETCS algorithm assuming the maximum processor frequencies and the maximum link frequencies so that there is reasonable slack for energy reduction. All the approaches are implemented in Matlab version R2015a. We use fmincon, quadprog and intlinprog solvers to solve the NLP, quadratic programming and ILP problems, respectively. The hardware platform consists of Intel(R) Core(TM) i5-4570 CPU with a clock frequency of 3.20 GHz, 8.00 GB memory, and 3 MB caches.

6.2 Results and Discussion

Experiments with Conditional Task Graphs. In the first set of experiments we choose eight benchmarks and their details are given in Table 6 where x/y/z/D stands for the number of tasks, the number of OR-FORK tasks, the number of conditions and the deadline of the application in seconds, respectively. The column with heading Dim represents NoC dimensions. The benchmarks in Table 6 are the same benchmarks used in [24].

Figure 8(a) shows that IOETCS-ILP achieves an average improvement of 31%, a maximum improvement of 62% for CTG 7 and a minimum improvement of 1.03% for CTG 1 over ILP-vpv-flv. It achieves an average improvement of 27%, a maximum improvement of 61% for CTG 3 and a minimum improvement of 7.9% for CTG 6 over ILP-fpv-vlv. IOETCS-Heuristic achieves an average improvement of 23%, a maximum improvement of 40% for CTG 5 and a minimum improvement of 1.3% for CTG 1 in comparison to ILP-vpv-flv. It achieves an average improvement of 18%, a maximum improvement of 61% for CTG 3 and a minimum improvement of 4% for CTG 6 over ILP-fpv-vlv. We observe that ILP-vpv-flv performs significantly better in terms of energy consumption if the computation energy dominates the total energy, and ILP-fpv-vlv performs better if communication energy dominates the total energy. CTGs 5, 6 and 7 favour ILP-fpv-vlv as the communication volumes for these benchmarks are significantly larger than the execution times of task nodes. Both IOETCS-ILP and IOETCS-Heuristic distribute slacks efficiently between communication nodes and task nodes and thus perform significantly better than ILP-vpv-flv and ILP-fpv-vlv. In terms of running time both ILP-vpv-flv and ILP-fpv-vlv run slightly faster than IOETC-SILP and IOETCS-Heuristic. This is because the search space of ILP-vpv-flv and ILP-fpv-vlv is smaller as compared to IOETCS-ILP and IOETCS-Heuristic. ILP-vpv-flv only scales processor voltages and ILP-fpv-vlv only scales link voltages.

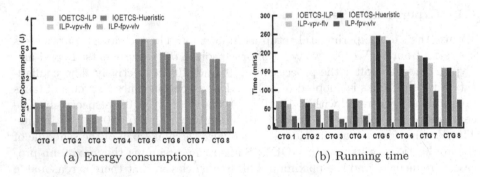

(a) Energy consumption   (b) Running time

Fig. 8. Comparison of eight benchmarks in Table 6

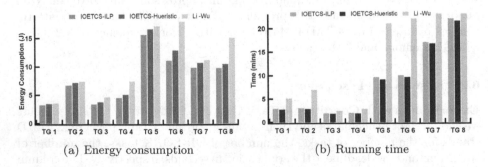

(a) Energy consumption   (b) Running time

Fig. 9. Comparison of eight benchmarks in Table 5

Whereas, IOETCS-ILP and IOETCS-Heuristic scale both the processor voltages and the link voltages. We choose two real-world benchmarks vehicle cruise controller [30] and Robot control [1] that are the task graphs of actual applications. These benchmarks are executed on 3 × 3 NoC where the processors are selected randomly as either Type 1 or Type 2. As shown in Fig. 10(a) IOETCS-ILP and IOETCS-Heuristic perform significantly better than ILP-vpv-flv and ILPfpv- vlv in terms of energy consumption. In terms of running time IOETCS-ILP and IOETCS-Heuristic take longer compared to ILP-vpv-flv and ILP-fpv-vlv as shown in Fig. 10(b). The reason is that IOETCS algorithm cannot find a feasible solution for some sub-problems, and thus the solver takes a longer time to converge.

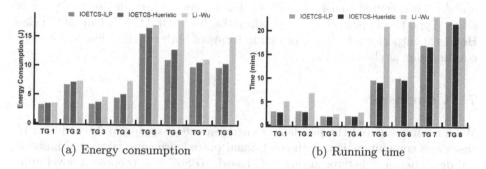

(a) Energy consumption   (b) Running time

Fig. 10. Comparison of real-world benchmarks with conditional precedence constraints

Experiments with Non-conditional Task Graphs. Our approach is the first approach that solves the problem discussed in this paper for CTG model. Since task model with unconditional precedence constraints is a special case of CTG we compare the performance of our approaches with the state-of-the-art approach for task model with unconditional precedence constraints. To demonstrate the effectiveness of our approach on task graphs without conditional precedence constraints, we have conducted a second set of experiments. We choose eight task graphs (TG) and their details are given in Table 5 where a/b/D stand for the number of tasks, the number of edges and the deadline of the application in seconds, respectively. The column with the heading Dim represents NoC dimensions.

Figure 9(a) gives a comparison of 8 benchmarks in Table 5 in terms of energy consumption where all the processors are of Type 1. IOETCS-ILP achieves an average improvement of 31%, a maximum improvement of 61% for TG 6 and a minimum improvement of 9% for TG 1 over Li-Wu approach. IOETCS-Heuristic achieves an average improvement of 20%, a maximum improvement of 46% for TG 4 and a minimum improvement of 2% for TG 1 over Li-Wu approach. We observe that Li-Wu approach makes very poor mapping decisions for heterogeneous processors. The benchmarks TG 3, TG 4, TG 6 and TG 8 are executed on MPSoCs where the processors are randomly selected as either Type 1 or Type 2. The rea-

son for poor performance of Li-Wu approach is that it does not take into account
the energy profiles of processors when making mapping decisions.

The benchmarks TG 1, TG 2, TG 5 and TG 7 are executed on MPSoCs with
homogeneous processors (Type 1). As a result, Li-Wu approach performs consid-
erably better for these benchmarks than for other benchmarks.

As shown in Fig. 9(b), in terms of running time, IOETCS-ILP and IOETCS-
Heuristic run approximately three times faster than Li-Wu approach. The major
reason is that the genetic algorithm takes significantly longer time as it constructs
a new schedule for each candidate solution using ETFGBF.

We have chosen two real-world benchmarks JPEG encoder [18] and Automatic
Target Recognition (ATR) [22]. JPEG encoder is executed on a 3×3 MPSoC
and ATR is executed on a 4×5 MPSoC. The processors are randomly selected as
either Type 1 or Type 2. For both benchmarks, both IOETCS-ILP and IOETCS-
Heuristic outperform Li-Wu approach in terms of both running time and energy
consumption as shown in Fig. 7.

7   Conclusion

We have investigated the problem of energy-aware mapping and scheduling of
tasks and communications with conditional precedence constraints and individual deadlines on a heterogeneous NoC-based MPSoC and propose a novel approach. Our approach reduces the total expected energy consumption by collectively optimizing the voltages/frequencies of processors and NoC links. The
IOETCS algorithm maps tasks to processors and serializes communications that
use same communication links. It constructs a unified schedule and assigns volt-
ages/frequencies to tasks and communications collectively assuming continuous voltages/frequencies. The IOETCS algorithm significantly narrows down the
search space for our ILP-based algorithm and our heuristic for assigning discrete
frequencies/voltages to tasks and communications. The experimental results show
that in terms of energy consumption, our approach using either ILP or heuristic
outperforms the state-of-the-art approach proposed by Li and Wu [22] that considers only unconditional task graphs. Compared to the stateof- the-art approach,
our ILP-based approach achieves an average improvement of 31%, a maximum
improvement of 61% and a minimum improvement of 9%, and our heuristic-based
approach achieves an average improvement of 20%, a maximum improvement of
46% and a minimum improvement of 2%. In terms of running time, our approach
is approximately 3 times faster than the state-of-the- art approach.

References

1. Standard task graph. http://www.kasahara.elec.waseda.ac.jp. Accessed 4 Sept 2017
2. Andrei, A., Eles, P., Jovanovic, O., Schmitz, M., Ogniewski, J., Peng, Z.: Quasi-static voltage scaling for energy minimization with time constraints. IEEE Trans. Very Large Scale Integr. (VLSI) Syst. **19**(1), 10–23 (2011)
3. Andrei, A., Eles, P., Peng, Z., Schmitz, M.T., Al Hashimi, B.M.: Energy optimization of multiprocessor systems on chip by voltage selection. IEEE Trans. Very Large Scale Integr. (VLSI) Syst. **15**(3), 262–275 (2007)
4. Andrei, A., Schmitz, M., Eles, P., Peng, Z., Al Hashimi, B.M.: Simultaneous communication and processor voltage scaling for dynamic and leakage energy reduction in time-constrained systems. In: Proceedings of the 2004 IEEE/ACM International Conference on Computer-Aided Design, pp. 362–369. IEEE Computer Society (2004)
5. Andrei, A., Schmitz, M., Eles, P., Peng, Z., Al-Hashimi, B.M.: Overhead-conscious voltage selection for dynamic and leakage energy reduction of time-constrained systems. IEE Proc. Comput. Digit. Tech. **152**(1), 28–38 (2005)
6. Bambagini, M., Marinoni, M., Aydin, H., Buttazzo, G.: Energyaware scheduling for real-time systems: a survey. ACM Trans. Embed. Comput. Syst. (TECS) **15**(1), 7 (2016)
7. Burd, T.D., Brodersen, R.W.: Energy efficient CMOS microprocessor design. In: Proceedings of the Twenty-Eighth Hawaii International Conference on System Sciences, vol. 1, pp. 288–297. IEEE (1995)
8. Cai, Y., Schmitz, M.T., Al-Hashimi, B.M., Reddy, S.M.: Workload-ahead-driven online energy minimization techniques for battery-powered embedded systems with time-constraints. ACM Trans. Des. Autom. Electron. Syst. (TODAES) **12**(1), 5 (2007)
9. Chen, G., Huang, K., Knoll, A.: Energy optimization for real-time multiprocessor system-on-chip with optimal DVFS and DPM combination. ACM Trans. Embed. Comput. Syst. (TECS) **13**(3s), 111 (2014)
10. Choi, K., Soma, R., Pedram, M.: Fine-grained dynamic voltage and frequency scaling for precise energy and performance tradeoff based on the ratio of off-chip access to on-chip computation times. IEEE Trans. Comput. Aided Des. Integr. Circuits Syst. **24**(1), 18–28 (2005)
11. DeNero, J., Klein, D.: The complexity of phrase alignment problems. In: Proceedings of the 46th Annual Meeting of the Association for Computational Linguistics on Human Language Technologies: Short Papers, pp. 25–28. Association for Computational Linguistics (2008)
12. Engel, M., Spinczyk, O.: A radical approach to network-on-chip operating systems. In: 42nd Hawaii International Conference on System Sciences, HICSS 2009, pp. 1–10. IEEE (2009)
13. Ge, Y., Zhang, Y., Malani, P., Qing, W., Qiu, Q.: Low power task scheduling and mapping for applications with conditional branches on heterogeneous multiprocessor system. J. Low Power Electron. **8**(5), 535–551 (2012)
14. Gebotys, C.H., Gebotys, R.J.: Power minimization in heterogeneous processing. In: Proceedings of the Twenty-Ninth Hawaii International Conference on System Sciences, vol. 1, pp. 330–337. IEEE (1996)

15. Ghosh, P., Sen, A., Hall, A.: Energy efficient application mapping to NoC processing elements operating at multiple voltage levels. In: Proceedings of the 2009 3rd ACM/IEEE International Symposium on Networks-on-Chip, pp. 80–85. IEEE Computer Society (2009)

16. Goh, L.K., Veeravalli, B., Viswanathan, S.: Design of fast and efficient energy-aware gradient-based scheduling algorithms heterogeneous embedded multiprocessor systems. IEEE Trans. Parallel Distrib. Syst. **20**(1), 1–12 (2009)

17. Huang, J., Buckl, C., Raabe, A., Knoll, A.: Energy-aware task allocation for network-on-chip based heterogeneous multiprocessor systems. In: 19th Euromicro International Conference on Parallel, Distributed and Network-Based Processing (PDP), pp. 447–454. IEEE (2011)

18. In, J., Shirani, S., Kossentini, F.: JPEG compliant efficient progressive image coding. In: Proceedings of the IEEE International Conference on Acoustics, Speech and Signal Processing, vol. 5, pp. 2633–2636 (1998)

19. Kang, J., Ranka, S.: Dynamic slack allocation algorithms for energy minimization on parallel machines. J. Parallel Distrib. Comput. **70**(5), 417–430 (2010)

20. Lee, H.G., Chang, N., Ogras, U.Y., Marculescu, R.: On-chip communication architecture exploration: a quantitative evaluation of point-to- point, bus, and network-on-chip approaches. ACM Trans. Des. Autom. Electron. Syst. (TODAES) **12**(3), 23 (2007)

21. Leung, L.-F., Tsui, C.-Y., Ki, W.-H.: Minimizing energy consumption of multiple-processors-core systems with simultaneous task allocation, scheduling and voltage assignment. In: Proceedings of the Asia and South Pacific Design Automation Conference, ASP-DAC 2004, pp. 647–652. IEEE (2004)

22. Li, D., Jie, W.: Energy-efficient contention-aware application mapping and scheduling on NoC-based mpsocs. J. Parallel Distrib. Comput. **96**, 1–11 (2016)

23. Li, K.: Power and performance management for parallel computations in clouds and data centers. J. Comput. Syst. Sci. **82**(2), 174–190 (2016)

24. Lombardi, M., Milano, M., Ruggiero, M., Benini, L.: Stochastic allocation and scheduling for conditional task graphs in multi-processor systems-on-chip. J. Sched. **13**(4), 315–345 (2010)

25. Lu, Z.: Using wormhole switching for Networks on Chip: feasibility analysis and microarchitecture adaptation. Ph.D. thesis, KTH (2005)

26. Marcon, C., Calazans, N., Moraes, F., Susin, A., Reis, I., Hessel, F.: Exploring NoC mapping strategies: an energy and timing aware technique. In: Proceedings of the conference on Design, Automation and Test in Europe, vol. 1, pp. 502–507. IEEE Computer Society (2005)

27. Mei, J., Li, K.: Energy-aware scheduling algorithm with duplication on heterogeneous computing systems. In: Proceedings of the 2012 ACM/IEEE 13th International Conference on Grid Computing, pp. 122–129. IEEE Computer Society (2012)

28. Mishra, R., Rastogi, N., Zhu, D., Mosse, D., Melhem, R.: Energy aware scheduling for distributed real-time systems. In: Proceedings of International Conference on Parallel and Distributed Processing Symposium, p. 21.2 (2003)

29. Mittal, S.: A survey of techniques for improving energy efficiency in embedded computing systems. Int. J. Comput. Aided Eng. Technol. **6**(4), 440–459 (2014)

30. Pop, P.: Scheduling and communication synthesis for distributed real-time systems. Department of Computer and Information Science, Linköpings universitet (2000)

31. Schmitz, M.T., Al-Hashimi, B.M., Eles, P.: Iterative schedule optimization for voltage scalable distributed embedded systems. ACM Trans. Embed. Comput. Syst. (TECS) **3**(1), 182–217 (2004)

32. Shin, D., Kim, J.: Communication power optimization for networkon- chip architectures. J. Low Power Electron. **2**(2), 165–176 (2006)
33. Singh, J., Betha, S., Mangipudi, B., Auluck, N.: Contention aware energy efficient scheduling on heterogeneous multiprocessors. IEEE Trans. Parallel Distrib. Syst. **26**(5), 1251–1264 (2015)
34. Sen, S., Huang, Q., Li, J., Cheng, X., Peng, X., Shuang, K.: Enhanced energy-efficient scheduling for parallel tasks using partial optimal slacking. Comput. J. **58**(2), 246–257 (2014)
35. Tariq, U.U., Wu, H.: Energy-aware scheduling of conditional task graphs with deadlines on MPSoCs. In: IEEE 34th International Conference on Computer Design (ICCD), pp. 265–272. IEEE (2016)
36. Topcuoglu, H., Hariri, S., Min-you, W.: Performance-effective and lowcomplexity task scheduling for heterogeneous computing. IEEE Trans. Parallel Distrib. Syst. **13**(3), 260–274 (2002)
37. Ye, T.T., De Micheli, G., Benini, L.: Analysis of power consumption on switch fabrics in network routers. In: Proceedings of the 39th Annual Design Automation Conference, pp. 524–529. ACM (2002)
38. Nemirovskii, A., Nesterov, Y.: Interior Point Polynomial Algorithms in Convex Programming. SIAM, Philadelphia (1987)
39. Zhuravlev, S., Saez, J.C., Blagodurov, S., Fedorova, A., Prieto, M.: Survey of energy-cognizant scheduling techniques. IEEE Trans. Parallel Distrib. Syst. **24**(7), 1447–1464 (2013)
40. Zong, Z., Manzanares, A., Ruan, X., Qin, X.: EAD and PEBD: two energy-aware duplication scheduling algorithms for parallel tasks on homogeneous clusters. IEEE Trans. Comput. **60**(3), 360–374 (2011)
41. Zong, Z., Qin, X., Ruan, X., Bellam, K., Nijim, M., Alghamdi, M.: Energy-efficient scheduling for parallel applications running on heterogeneous clusters. In: Proceedings of International Conference on Parallel Processing, p. 19. IEEE (2007)

Author Index

Printed in the United States
By Bookmasters

Printed in the United States
By Bookmasters